Multifunctional Ferroelectric Materials

Edited by Dipti Ranjan Sahu

Published in London, United Kingdom

IntechOpen

Supporting open minds since 2005

Multifunctional Ferroelectric Materials
http://dx.doi.org/10.5772/intechopen.92527
Edited by Dipti Ranjan Sahu

Contributors
Veronica Lucero Villegas Rueda, Mohamed Mansour, Haruichi Kanaya, B. Chandra Sekhar, B. Dhanalakshmi, B.Srinivasa Rao, S. Ramesh, K.Venkata Prasad, P.S.V Subba Rao, B.Parvatheeswara Rao, Maheswar Panda, Manoranjan Kar, Lagen Kumar Pradhan, paramjit kour, Sudipta Kishore Pradhan, Muzaffar Iqbal Khan, Trilok Chandra Upadhyay, Dipti Ranjan Sahu

First published in London, United Kingdom, 2021 by IntechOpen
IntechOpen is the global imprint of INTECHOPEN LIMITED, registered in England and Wales, registration number: 11086078, 5 Princes Gate Court, London, SW7 2QJ, United Kingdom
Printed in Croatia

British Library Cataloguing-in-Publication Data
A catalogue record for this book is available from the British Library

Additional hard and PDF copies can be obtained from orders@intechopen.com

Multifunctional Ferroelectric Materials
Edited by Dipti Ranjan Sahu
p. cm.
Print ISBN 978-1-83968-991-8
Online ISBN 978-1-83968-992-5
eBook (PDF) ISBN 978-1-83968-993-2

For EU product safety concerns:
IN TECH d.o.o., Prolaz Marije Krucifikse Kozulić 3, 51000 Rijeka, Croatia, info@intechopen.com or visit our website at intechopen.com.

Meet the editor

Dr. Dipti Ranjan Sahu is Associate Professor of Physics, Department of Natural and Applied Sciences, Namibia University of Science and Technology (NUST). He received a Ph.D. in Physics from the Institute of Materials Science, Utkal University, India. He has worked as a postdoctoral researcher and visiting scientist at several institutions, including National Taiwan University, National Cheng Kung University, Taiwan, and the University of Witwatersrand, South Africa. His research focuses on multifunctional materials including nanomaterials, ceramics, composites, spintronics, ferroelectrics, and magnetic materials, and the application of these functional materials in devices. He has published more than 101 peer-reviewed articles and more than 104 research articles in conference proceedings and meetings. He has also published two books and two book chapters.

Contents

Preface

Ferroelectricity is a well-known phenomenon commonly used in scientific and industrial communities. Ferroelectrics are spontaneously polarized materials that exhibit a range of properties and applications. Such materials are classified based on their behaviour, properties, functions, and applications.

New ferroelectrics and hybrid structures are developed using different synthesis and fabrication techniques. Different phenomena and interesting properties are studied theoretically and experimentally using advanced characterization techniques. Ferroelectric materials are the building blocks of different devices and technological innovations.

This book presents an overview of the basic phenomenon of ferroelectricity as well as different ferroelectrics and ferroelectric devices, including their theoretical study, synthesis, characterization, and application. It provides information for students, researchers, and professionals. It addresses scientific advances in the field of ferroelectricity for the study of next-generation devices. Understanding the fundamentals of ferroelectrics and their related properties such as size effects, phase transition, and symmetry can give new and fascinating physics for the development of new exotic materials.

This book contains eight chapters presenting some of the most exciting theoretical concepts and developments of ferroelectric materials. Chapters 1 and 2 introduce ferroelectrics and the basic properties of ferroelectric materials. Chapters 3 and 4 deal with different types of ferroelectrics such as perovskite ferroelectrics and relaxor ferroelectrics. Chapter 5 examines piezoelectricity and its application. Chapters 6–8 discuss the properties and basic performance of ferroelectric materials. These chapters explain the fundamentals as well as present research on ferroelectric materials for technological innovation.

I would like to thank the chapter authors for their expert contributions. I would also like to thank the staff at IntechOpen who helped throughout the preparation and publication of this book.

Dipti Ranjan Sahu
Faculty of Health and Applied Sciences,
Department of Natural and Applied Sciences,
Namibia University of Science and Technology,
Windhoek, Namibia

Chapter 1

Introductory Chapter: Basics and Importance of Ferroelectric Materials for Applications

Dipti Ranjan Sahu

1. Introduction

Multifunctional advanced materials are the building block of new device for different application. In recent years research on different fields ranging from cross disciplinarity to multidisciplinary fields including nanotechnology, materials science, physical and biological, increases rapidly based on the growth and development off new and novel materials. Concept of designing and fabricating new and exotic multifunctional materials relating to electronic, spintronics and magnetoelectric dated back to the original work [1, 2] of James Clerks Maxwell (1865) and M. Curie (1894). Among these materials, ferroelectric materials have got its own special technological status for its intriguing properties due to charge and exploiting the coupling of charge with other basic material parameter through electric and magnetic field which make these materials very much promising for future generation devices based on electronics, photonics, opto-electronics and spintronics, etc.

1.1 Definition of ferroelectrics and its basics

Ferroelectrics are known by its stable polarization state [3, 4]. There is a hysteresis effect associated with ferroelectric materials [5–7]. Different type of interactions between electrical, thermal and mechanical properties shown by ferroelectric materials has wealth of functionalities. Ferroelectric materials behavior and properties changes with temperature, electric field, pressure, strain, etc. which make variety of application in different devices.

In addition, some basic requirements of a ferroelectric materials are enumerated as follows:

1.1.1 Symmetry

Ferroelectricity requires the condition of non-centro symmetric and allows an electric polarization [8].

1.1.2 Electrical properties

Ferroelectrics shows spontaneous electric polarization, which can switch direction with an applied electric field.

1.1.3 Chemistry of "d^0-ness"

The ferroelectric materials with ABO_3-type perovskite structure have d^0' electron configuration on the B-site cation [8].

IntechOpen

1.1.4 Structural anisotropy and distortions

The structural anisotropy responsible for the ferroelectricity due to position of different ions in the crystal lattice [9]. In some cases, the distortion creates in materials due to the position of the certain 'd' orbital occupancies cations shows ferroelectricity [10]. These materials produce a ferroelectric state in a magnetically ordered state.

1.2 Types of ferroelectric materials

Depending on the mechanism of ferroelectricity, one can as well as single out four major types of ferroelectrics, although there are certainly others [11], which are ferroelectricity by mixed perovskite, ferroelectricity by ordering of lone pair d^0 ions, ferroelectricity by charge ordered system, ferroelectricity by geometric arrangement of atoms and shown in (**Figure 1**).

In addition to above four types of ferroelectric material, there are other types of ferroelectrics.

1.2.1 Spiral type ferroelectrics

Specific magnetic spiral is responsible for ferroelectric (Ex. $TbMnO_3$, $Ni_3V_2O_6$, and $MnWO_4$) [11].

Collinear Magnetic Structures ferroelectrics: In this materials ferroelectricity appears in collinear magnetic structure. These ferroelectrics are generally known as magnetically driven ferroelectrics [12]. The simplest example is Ca_3CoMnO_6.

1.3 Key challenges of ferroelectic materials

In order to bring these materials for applications, one has to control the state of the material system by phase control [13–15] or by domain control [16–18]. It is also evident from the literature that several theory and models are proposed to create

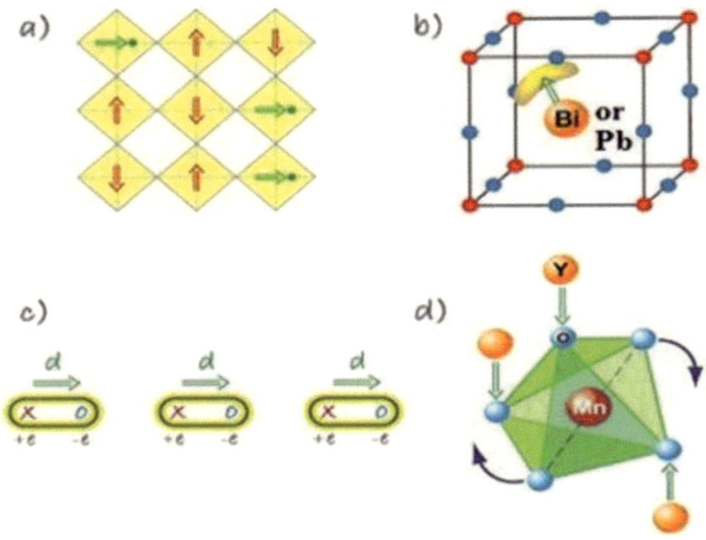

Figure 1.
(a) ferroelectricity by mixed perovskite, (b) ferroelectricity by ordering of lone pair do ions, (c) ferroelectricity by charge ordered system, and (d) ferroelectricity by geometric arrangement of atoms. Figure adopted from Khomskii D. [11].

magnetically-induced ferroelectricity for electronic applications which are generally caused by the presence of spin–orbit interaction [19, 20]. In some cases, spin–orbit coupling is not a requisite [21, 22] rather ferroelectric distortion can cause a small magnetization which can apply to spintronic devices.

For long time there is active research going on ferroelectric materials, still the exact nature and value of spontaneous polarization for application in some of the ferroelectric materials is the matter of long-standing discussion [22]. On the other hand, the mechanism of ferroelectric phase transitions in some ferroelectric materials is not understood clearly and information related to its properties are also not available [23]. Therefore, it is the need of the hour to study and develop novel ferroelectric material for applications.

1.4 Application of ferroelectric materials

Ferroelectric materials show interesting behavior and properties which can be manipulated for design and development of different types of devices. These materials can be used in information storage, spintronic, computing, communications, memories, actutators, motors and sensors [24, 25]. Ferroelectric materials can be used as insulators and semiconductor process integration materials in industry [26–29]. Recent success studies on ferroelectric materials for integrated circuits creates great interest in researcher for development of new ferroelectric materials with different functionalities [30] and address the problem associated with the materials for applications. Still there is plenty of opportunities ahead for ferroelectric material for wide variety of applications.

Acknowledgements

Author highly acknowledges the financial and admirative support of department of Natural and Applied Sciences, Namibia University of Science and Technology, Windhoek. Namibia for this work.

Author details

Dipti Ranjan Sahu
Department of Natural and Applied Sciences, Namibia University of Science and Technology, Windhoek, Namibia

*Address all correspondence to: dsahu@nust.na

References

[1] Curie, P., J. Phys. Theor. Appl. 1894;3:393-415.

[2] Maxwell, J. C., Phil. Trans. R. Soc. Lond. 1865;155:459-512.

[3] Fiebig, M., J. Phys. D: Appl. Phys. 2005;38:R123.

[4] Mitsui, T., "Ferroelectrics and antiferroelectrics," in Springer Handbook of Condensed Matter and Materials Data, edited by W. Martienssen and H. Warlimont (Springer, Heidelberg, 2005), pp. 903-938.

[5] Xu, Y., Ferroelectric Materials and their Applications (Elsevier, 1991).

[6] Trolier-McKinstry, S., Am. Ceram. Soc. Bull. 2020;99(1):22-23.

[7] Mikolajick, T., Schroeder, U., Slesazeck, S., IEEE Trans. Electron Devices 2020;67(4):1434-1443.

[8] Hill, N. A., J. Phys. Chem. B 2000;104:6694.

[9] Schmid, H., Int. J. Magn. 1973;4:337.

[10] Mizokawa, T., Khomskii, D. I., Sawatzky, G. A., Phys. Rev. B 1999;60:7309.

[11] Khomskii, D., Physics 2009;2:20.

[12] Choi, Y. J., Phys. Rev. Lett. 2008;100:047601.

[13] Kimura, T., Goto, T., Shintani, H., Ishizaka, K., Arima, T., Tokura, T., Nature 2003;426:55-58.

[14] Lottermoser, T., Lonkai, T., Amann, U., Hohlwein, D., Ihringer, J., Fiebig, M., Nature (London) 2004;430: 541-544.

[15] Hur, H., Park, S., Sharma, P. A., Ahn, J. S., Guha, S., Cheong, S. W., Nature 2004;429: 392-395.

[16] Fiebig, M., Lottermoser, T., Frohlich, D., Goltsev, A. V., Pisarev, R. V., Nature (London) 2002;419:818-820.

[17] Zhao, T., Scholl, A., Zavaliche, F., Ramesh, R., Nature Materials 2008;5(10):823-839.

[18] Katsura, H., Nagaosa, N., Balatsky, A. V., Phys. Rev. Lett. 2005;95:057205.

[19] Sergienko, I. A., Dagotto, E., Phys. Rev. B 2006;73:094434.

[20] Sergienko, I. A., Sen, C., Dagotto, E., Phys. Rev. B 2006;97:227204.

[21] Picozzi, S., Yamauchi, K., Sanyal, B., Sergienko, I. A., Dagotto, E., Phys. Rev. Lett. 2007;99:227201.

[22] Neaton, J. B., Ederer, C., Waghmare, U. V., Spaldin, N. A., Rabe, K. M., Phys. Rev. B 2005;71:014113.

[23] Vasudevan, R. K., Balke, N., Maksymovych., Jesse, S., Kalinin, S. V. Appl. Phys. Revs 2017;4:021302.

[24] MacDonald, A. H., Schiffer, P., Samarth, M., Nature Materials 2005;4:195-202.

[25] Coey, J. M. D., Venkatesan, M., J. Appl. Phys., 2002;91:8345-8350.

[26] Moodera, J. S., Kinder, L. R., Wong, T. M., Meservey, R., Phys. Rev. Lett. 1995;74:3273-3276.

[27] Zhang, X. Y., Lai, C. W., Zhao, X., Wang, D. Y., Dai, J. Y., Appl. Phys. Lett. 2005;87:143102(1-3).

[28] Wang, Y. P., Zhou, L., Zhang, M. F., Chen, X. Y., Liu, J. M., Liu, Z. G. Appl. Phys. Lett. 2004;84(10):1731-1733.

[29] Kim, T. Y., Kim, S. K., Kim, S. W., Nano Convergence 2018;5:30(1-16).

[30] Mikolajick, T., Slesazeck., Mulaosmanovic, H., Park, M. H., Fichtner, S., Lomenzo, P. D., Hoffmann, J. Appl. Physics 2021;129:1009011-21.

Chapter 2

General Introduction to Ferroelectrics

Muzaffar Iqbal Khan and Trilok Chandra Upadhyay

Abstract

In this chapter "General introduction to ferroelectrics" we contribute the basic idea of the fundamental properties of ferroelectrics. We focus on the following properties in the chapter such as basic introduction, classification, ferroelectric phase transitions, spontaneous polarization, local field, dielectric properties, polarizability, thermodynamics of ferroelectricity and applications of ferroelectrics. Ferroelectric materials are unusual dielectric which possesses reversible spontaneous electric polarization which can be reversed by application of stress or electric field which exhibit a range of properties. These properties are widely used in the today's scientific and industrial technology. The large number of areas due to their peculiar and interesting properties such as high permittivity capacitors, ferroelectric non-volatile FeRAM memories, pyroelectric sensors, piezoelectric and transducers, electrooptic and optoelectronic devices, etc.

Keywords: dielectrics, ferroelectrics, polarization, piezoelectric, pyroelectric, hysteresis loop, phase transitions

1. Introduction

The investigations of dielectrics, ferroelectrics, sensor, dipolar glasses and composite materials have attracted great attention in solid-state physics and material science in recent years. These studies reveal inter-and intra-molecular interactions and encourage increasing applications of these materials in modern technology. Dielectric substances are insulators or poor conductor of electricity. In these materials, the electrostatic field persists for a long time. These materials do not have free electrons, but the application of the electric field changes their behaviour. They have the ability to be polarized under the action of the electric field [1]. Dielectric materials are classified into two main categories (i) Non-ferroelectric (called normal dielectric or paraelectric) materials and (ii) Ferroelectric materials. The non-ferroelectric materials are divided into three categories according to the prevailing polarization mechanism as (i) non-polar dielectrics, (ii) polar dielectrics and (iii) dipolar dielectrics [1]. The non-polar dielectric materials consist of one type of atoms. These types of dielectric materials become polarized in an external electric field due to the relative displacement of electric charge with respect to the nucleus. The polar dielectric materials are made up of molecules without a permanent dipole moment, and dipolar dielectrics include materials whose molecules possess a permanent dipole moment.

Ferroelectric materials are unusual dielectrics that possess reversible spontaneous electric polarization, which can be reversed by applying stress or electric field. This property of showing spontaneous polarization persists over a certain temperature interval [2]. Above the critical temperature, called Curie temperature or transition temperature, the substance loses its property of spontaneous polarization and becomes paraelectric. This change of phase property, i.e., ferroelectric phase to paraelectric phase, is associated with anomalous behaviour of many physical properties along with a change of crystal structure from low to a high symmetry. The alignment of the electric dipoles may extend only over a region of the crystal, while in another region, the direction of spontaneous polarization may be reversed. Such regions of uniform polarization are called domains, a term acquired from ferromagnetism. If we first applied a small uniform electric field (E), directed, (say) in the positive direction, we will induce uniform polarization P (a linear relationship between P and E) because the field is not larger enough to switch any of the domains with the unfavorable direction of polarization and the crystal will behave like the normal dielectric [2, 3].

A material can be either piezoelectric, pyroelectric or ferroelectric, only if its crystalline symmetry is inherent (i.e., it lacks an inversion centre). A basic principle due to Neumann is that any physical property exhibited by a crystal must have at least the symmetry of the point group of the crystal. Thus, the above properties, which are inherently asymmetric, can only arise in asymmetric crystals. All crystal structures can be divided into 32 crystal classes. Of the 32 crystallographic point groups, 11 exhibit centre symmetry, leaving 21 non-centrosymmetric point groups. One of the 21 groups, however, have an inversion centre, causing it to lose its non-centrosymmetric nature, leaving 20 non-centrosymmetric point groups which have asymmetric properties. All the crystals in these 20 classes are piezoelectric [3, 4]. When such a non-centrosymmetric crystal is subjected to mechanical stress, the ions are displaced from each other in an asymmetric manner, and the crystal becomes electrically polarized. This is called the piezoelectric effect. The inverse effect of it, i.e., an applied electric field produces strain (causes the material to either expand or contract, depending on the field direction) has also been observed. The piezoelectric effect is often used to convert electrical energy into mechanical energy and vice-versa. Quartz is the best example of piezoelectric material and the one most frequently used in transducers. Out of the 20 piezoelectric point groups, only 10 have a unique polar-axis responsible for the appearance of a spontaneous electric polarization even in the absence of an applied electric field [3, 4]. If a piezoelectric material also shows the change in spontaneous polarization (P_s) upon a change in temperature, according to the relation, $\Delta P_s = \lambda \Delta T$, where λ is the pyroelectric coefficient. This is called the pyroelectric effect. A pyroelectric material changes its unit cell dimensions up to temperature change. This causes the unit cell to either expand or contract, including a temperature-dependent polarization. The simplest example of a pyroelectric material is Wurtzite (hexagonal ZnS). In some pyroelectric materials, the spontaneous polarization can be reversed by an externally applied electric field, giving a dielectric hysteresis loop. Such materials are called ferroelectric materials, and the phenomenon of reversing the direction of polarity is called the ferroelectric effect. It is to be noted that both piezoelectric and pyroelectric are inherent properties of a material. On the other hand, ferroelectricity is an effect produced in a pyroelectric material by the application of an external electric field. Classification of these materials based on symmetry is shown schematically in **Figure 1**. The occurrence of ferroelectricity may be understood in terms of either (i) polarization catastrophe or (ii) transverse optical phonon mode [5, 6].

The relationship between the piezoelectric, pyroelectric and ferroelectric materials is shown schematically in **Figure 2**. A dielectric material is an electrical

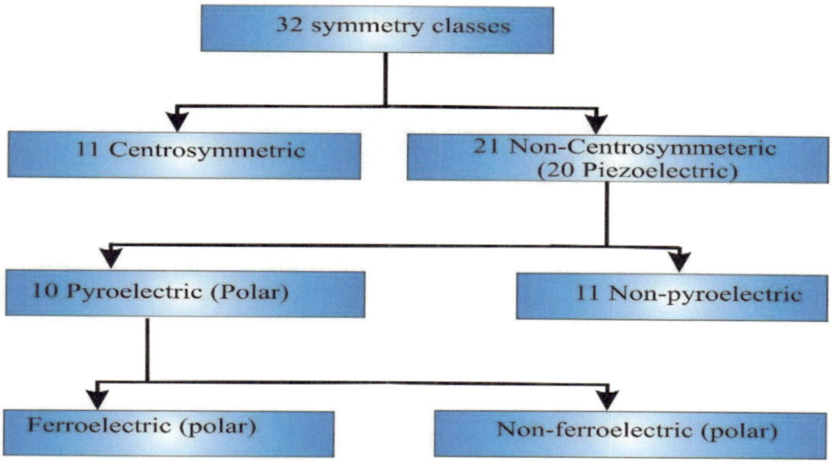

Figure 1.
Classification of piezoelectric and pyroelectric materials [6].

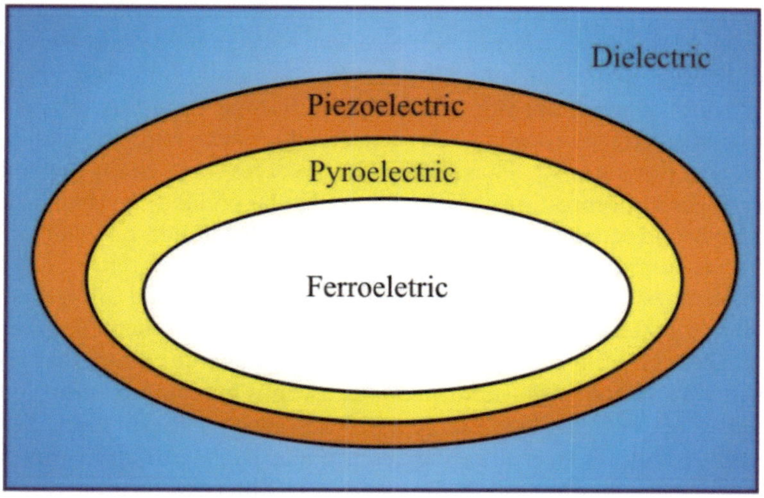

Figure 2.
Venn diagram showing the relationship between various types of dielectric materials [8].

insulator that can be polarized under an external applied electric field. Dielectric materials are often characterized by their dielectric permittivity, which describes the material's resistance against polarization by an external electric field [7, 8]. A group of dielectrics that show a change of strain or stress due to an applied external electric field or conversely to the change of the polarization due to a mechanical excitation are called piezoelectrics. Pyroelectrics are a group of piezoelectrics that show a change of polarization due to a change in temperature. Ferroelectric materials have both pyroelectric and piezoelectric properties.

Ferroelectricity is the phenomenon that refers to the state of spontaneous polarization, usually vanishes above a certain temperature called Curie or transition temperature (T_c). At the T_c, the crystal undergoes a phase transition from the polar state to the non-polar state. Three well-known characteristics of ferroelectrics are (i) their reversible polarization, (ii) their anomalous properties and (iii) their non-linearities. Above the transition temperature (T_c), the crystal is said to be the paraelectric state. The term paraelectric is analogous with paramagnetism; similarly, there is usually a rapid drop in the dielectric constant (ε) as the temperature increases [5–9]. In ferroelectrics, the temperature dependence of dielectric constant above the Curie point can be explained by the simple law called Curie–Weiss law

$$\varepsilon = \varepsilon_0 + \frac{C}{(T - T_0)} \tag{1}$$

where ε_0 is the part of dielectric constant independent of temperature, C is the Curie constant and T_0 the Curie–Weiss temperature and it is different from the Curie point T_c (contributed by electronic polarization). The phase transition at T_c can be first-order or of the higher (second) order. In the case of a first-order phase transition, $T_0 < T_c$, while for the second-order phase transition $T_0 = T_c$. Usually, the temperature-independent term ε_0 can be neglected since it is much smaller than the term $\frac{C}{(T-T_0)}$ when T is near T_0.

2. Ferroelectric phase transitions

A phase transition is the transformation of the thermodynamic system from one phase or state of matter to another. It is a collective phenomenon in which critical behaviour depends on a small number of parameters and is universal for many systems. During a phase transition of a given medium, certain properties of the medium change, often discontinuously, as a result of some external condition, such as temperature, pressure, etc. Phase transition involves some change of symmetry. According to Paul Ehrenfest, phase transitions can be divided into two groups known as first- and second-order phase transitions, depending on whether the transition is discontinuous or continuous, respectively. Paul Ehrenfest classified phase transitions based on the behaviour of the thermodynamic Gibbs free energy as a function of other thermodynamic variables. Under this scheme, phase transitions were labeled by the lowest derivative of the Gibbs free energy that is discontinuous at the transition. First-order phase transitions exhibit a discontinuity in the first derivative of the Gibbs free energy with respect to the thermodynamic variable [10]. Second-order phase transitions are continuous in the first derivative but exhibit discontinuity in a second derivative of the Gibbs free energy with respect to a thermodynamic variable [10]. In the first-order phase transition, volume, entropy and polarization of the crystal change discontinuously at the transition point. In the second-order phase transition, the specific heat changes discontinuously, ‘whereas volume, entropy and polarization change continuously at the phase transition point. In the first-order phase transition, the energy appearing as latent heat in an infinitely narrow temperature range interval, while in the second-order phase transition, there is no release of the latent heat but the expansion of the coefficient exhibits anomalous behaviour over a finite range of temperature [11, 12].

In ferroelectrics, two common types of phase transition are identified. These are named depending on how the order parameter (polarization) changes during the transition. It is common to observe that as the temperature is raised, the bulk polarization decreases and vanishes abruptly at a Curie temperature (T_c). This is a phase transition, just as in a ferromagnet raised above its Curie temperature or a solid raised above its melting point. It arises microscopically because as the temperature is raised, the thermal vibrations of the atoms in the solid cause fluctuations, which overcome the potential barrier between the two (or more) wells. For example, in a molecular crystal such as $NaNO_2$, where we imagine that each molecule can fluctuate between two configurations. Each of which has a double potential well, as shown in **Figure 3** and some interactions between the dipoles that tend to align them [13].

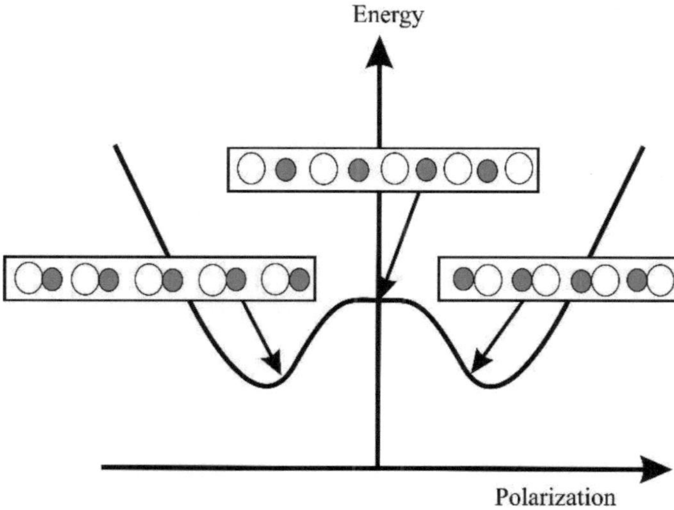

Figure 3.
Schematic potential well [13].

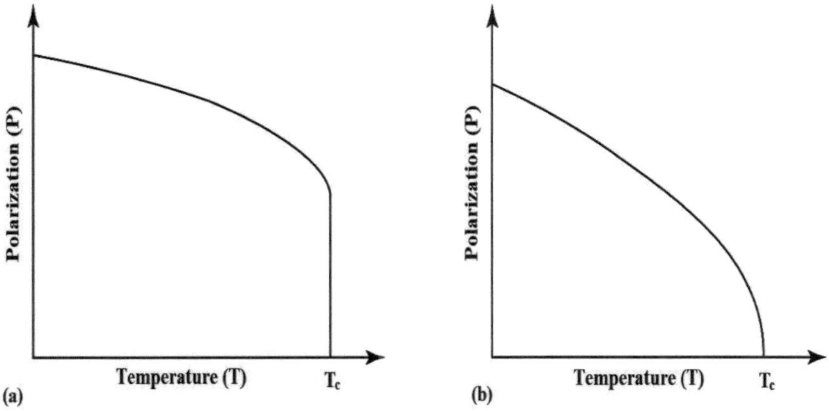

Figure 4.
Plots of spontaneous polarization versus temperature: (a) first-order transition (b) second order transition [13].

The detailed microscopic theory of how this happens will be different from material to material, but the macroscopic properties of the phase transition will be similar across many different classes of materials. A first-order phase transition is one that has a discontinuity in the order parameter itself, while a second-order phase transition is one that has a discontinuity in the first derivative of the order-parameter. In a first-order transition, the polarization varies continuously until the Curie temperature, at which there is a discontinuity shown in **Figure 4a**. In a second-order transition, the order parameter itself is a continuous function of temperature, but there is a discontinuity in its first derivative at Curie temperature shown in **Figure 4b** [13].

3. Spontaneous polarization

The intensity of polarization (P) is defined as the electric dipole moment per unit volume of the dielectric material. Spontaneous polarization (P_s) is a polarization that occurs under the influence of an internal process in a dielectric, without the effect of external factors. A ferroelectric crystal generally consists of regions called domains of homogeneous polarization, within each of which the polarization is in the same direction, but in the adjacent domains, the polarization is in different directions so that the net polarization of the specimen is equal to zero in the beginning when no electric field ($E = 0$) is applied. The polarization varies in a non-linear configuration with an electric field (E). This non-linear relation exhibits the closed curve called the hysteresis loop in the polarization when an electric field is applied, as shown in **Figure 5** [2–6]. When the electric field ($E = 0$) is zero, the spontaneous polarization (P_s) in a single domain specimen is either positive or negative in sign. As an applied electric field strength gradually increases in the direction of spontaneous polarization (P_s), the polarization (P) increases due to induced polarization such as electronic, ionic and dipolar types. As the electric field is increased further, more and more domains rotate along the direction of the electric field (E) until the polarization reaches a maximum value called the saturation value.

At this stage, the whole specimen represents a single domain. This is usually accompanied by a distortion in the crystal along the polarization direction. The extrapolation of the saturation value to zero field gives the magnitude of the spontaneous polarization (P_s). This value of P_s is the same as possessed by each domain before the application of the electric field. However, if the applied electric field decreases, the polarization also decreases but follows another path and does not become zero for zero electric field. The remaining polarization at this stage is called remnant polarization ($\pm P_r$) and the intercept on the E-axis, where P_r refers to the whole crystal block. In order to destroy the remnant polarization (P_r), the polarization of nearly half of the crystal is to be reversed by reversing the direction of the field, the electric field required to make the polarization zero is called the coercive field (E_c). Furthermore, an increase in the reverse field results in the saturation of polarization in the reverse direction. Reversing the electric field again, the hysteresis curve will be obtained. The relation between polarization (P) and applied electric field (E) is thus represented by a hysteresis loop (BDFGHB) which is the most important characteristic of the ferroelectric crystals. The most important feature of a ferroelectric is thus not the fact that it has a spontaneous polarization (P_s) but rather the fact that this spontaneous polarization can be reversed by means of an electric field [2–14].

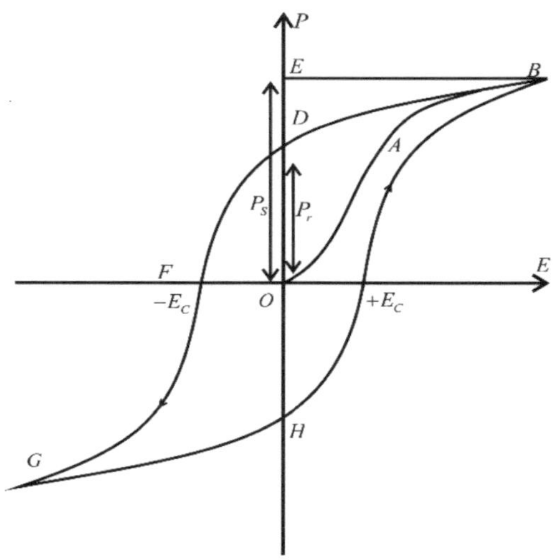

Figure 5.
Ferroelectric (P-E) hysteresis loop [13].

4. Dielectric properties

4.1 Local field

The electric field acting at the site of atom or molecule is, in general, significantly different from the macroscopic filed E and is known as the local field (E_{loc}). This field is responsible for the polarization of each atom or molecule of a material [5, 6]. For an atomic site with cubic crystal symmetry, the local field (E_{loc}) is expressed by the Lorentz-relation as

$$E_{loc} = E_0 + E_P + \frac{P}{3\varepsilon_0} = E + \frac{P}{3\varepsilon_0} \tag{2}$$

where $E = E_0 + E_P$. The field E_P is called the polarization field as it tends to oppose the external applied field E_0. Thus, apart from the macroscopic field (E), the local field also contains a term called Lorentz field (E_{loc}). The difference between the macroscopic field (E) and the Lorentz field (E_{loc}) may be understood as follows. The macroscopic field is macroscopic in nature, is an average value and is constant throughout the medium. On the other hand, the Lorentz field (E_{loc}) is a microscopic field and is periodic in nature. This is quite large at molecular sites representing that the molecules are more effectively polarized than they are under the average field.

4.2 Polarization and dielectric susceptibility

Generally, at ordinary electric fields, the magnitude of polarization (P) is directly proportional to the macroscopic electric field (E) at a given point of a dielectric [5–15]. It is expressed as

$$P \propto E \Rightarrow P = \varepsilon_0 \chi_e E \tag{3}$$

where ε_0 is the permittivity of free space and χ_e is the dielectric susceptibility. Thus, except for a constant factor (ε_0), the dielectric susceptibility is a measure of the polarization produced in the material per unit electric field. If the dielectric material slab is placed in a uniform electric field (E) with its normal parallel to the field. The dielectric displacement vector (D) for an isotropic or cubic medium relative to vacuum is defined in terms of the macroscopic field (E) as

$$D = \varepsilon_0 \varepsilon_r E = \varepsilon E = \varepsilon_0 E + P \tag{4}$$

where ε_r is called the relative permittivity or dielectric constant of medium and ε_0 is the permittivity or dielectric constant of free space, and P is the polarization. It is a scalar quantity for an isotropic medium and is always dimensionless. The dielectric constant (also called as permittivity of medium) is a measure of the degree to which a medium can resist the flow of charge, defined as the ratio of the dielectric displacement D to the macroscopic field intensity (E) as

$$\varepsilon_r = \frac{D}{\varepsilon_0 E} = \frac{\varepsilon_0 E + P}{\varepsilon_0 E} = 1 + \frac{P}{\varepsilon_0 E} = 1 + \chi_e \tag{5}$$

Eq. (5) gives the dielectric constant of an isotropic medium or cubic medium. This is represented by a scalar quantity. The dielectric susceptibility (χ_e) is related to the dielectric constant defined as

$$\chi_e = \frac{P}{\varepsilon_0 E} = \varepsilon_r - 1 \tag{6}$$

Thus, like susceptibility, the dielectric constant (ε_r) is also a measure of the polarization (P) of the material. The larger the polarization per unit resultant macroscopic field, the greater will be the dielectric constant of the dielectric medium. However, for anisotropic medium, the dielectric response ($\varepsilon_r$or χ_e) depends on the direction of the field and described by the components of the susceptibility tensor or of the dielectric tensor of the second rank

$$P_\mu = \chi_{\mu\nu}\varepsilon_0 E_\nu; \varepsilon_{\mu\nu} = \delta_{\mu\nu} + \chi_{\mu\nu} \tag{7}$$

4.3 Dielectric constant and polarizability

The polarizability (α) of an atom is defined in terms of the local electric field (Lorentz field) at the atom. The induced dipoles of moments (p) are proportional to the local field (E_{loc}) can be expressed as

$$p = \alpha E_{loc} \tag{8}$$

where α is known as the polarizability of an atom. For a non-spherical atom or isotropic medium, α will be a tensor quantity [6–16]. Thus, polarizability is an atomic property, whereas dielectric constant is a macroscopic property that depends upon the arrangement of atoms within the crystal. If all the atoms have the same polarizability (α) and there are N number of atoms per unit volume, the total polarization of the crystal may be expressed as the product of the polarizabilities of the atoms times the local field

$$P = \sum_j N_j p_j = \sum_j N_j \alpha_j E_{loc}(j) \tag{9}$$

where the summation is over all the atoms or atomic sites. N_j is the concentration and α_j is the polarizability of atom j and $E_{loc}(j)$ is the local field at atom sites j. For an isotropic dielectric medium, the local field given by the Lorentz relation Eq. (2) inside the crystal is everywhere the same so that it can be taken out of the summation sign from Eq. (9). Substituting the value of the local field from Eq. (2), the Eq. (9) becomes

$$P = \left(E + \frac{P}{3\varepsilon_0}\right)\sum_j N_j \alpha_j \tag{10}$$

On rearranging the terms and making use of Eq. (5) gives

$$\chi_e = \frac{P}{\varepsilon_0 E} = \frac{\sum_j N_j \alpha_j}{\left(\varepsilon_0 - \frac{1}{3}\sum_j N_j \alpha_j\right)} \tag{11}$$

Using Eq. (6), we get

$$\varepsilon_r = 1 + \frac{\frac{\sum_j N_j \alpha_j}{\varepsilon_0}}{1 - \frac{1}{3\varepsilon_0}\sum_j N_j \alpha_j} = \frac{1 + \frac{2}{3\varepsilon_0}\sum_j N_j \alpha_j}{1 - \frac{1}{3\varepsilon_0}\sum_j N_j \alpha_j} \tag{12}$$

Solving for $\sum_j N_j \alpha_j$ we get

$$\frac{\varepsilon_r - 1}{\varepsilon_r + 2} = \frac{1}{3\varepsilon_0} \sum_j N_j \alpha_j \tag{13}$$

This is known as the Clausius-Mossotti relation. It relates the dielectric constant to the atomic polarizability, but only for crystal structures for which the Lorentz field Eq. (2) obtains. The total polarization (α) can be expressed as the sum of three types of basic polarizability representing the most important contributions to the polarization [2–17] given as

$$\alpha = \alpha_e + \alpha_i + \alpha_d \tag{14}$$

where α_e, α_i and α_d are the electronic, ionic and dipolar polarizabilities, respectively shown in **Figure 6**.

i. Electronic polarizability (α_e): The electronic polarizability (α_e) arises due to the displacement of electrons in an atom relative to the atomic nucleus in the external electric field, as shown in **Figure 6a**. The polarization, as well as the dielectric constant of a material at optical frequencies, results mainly from the electronic polarizability (α_e). The optical range Eq. (13) reduces as

$$\frac{n^2 - 1}{n^2 + 2} = \frac{1}{3\varepsilon_0} \sum_j N_j \alpha_j (\text{electronic}) \tag{15}$$

where n being the refractive index which is related to the dielectric constant (ε_r) by the relation

$$n^2 = \varepsilon_r \tag{16}$$

ii. Ionic polarizability (α_i): The ionic polarizability (α_i) arises due to the relative displacement of positive and negative ions from their equilibrium positions to a distance less than the distance between adjacent ions. The cations are displaced parallel to the Lorentz field, and the anions are displaced in the opposite direction, as shown in **Figure 6b** [5–18].

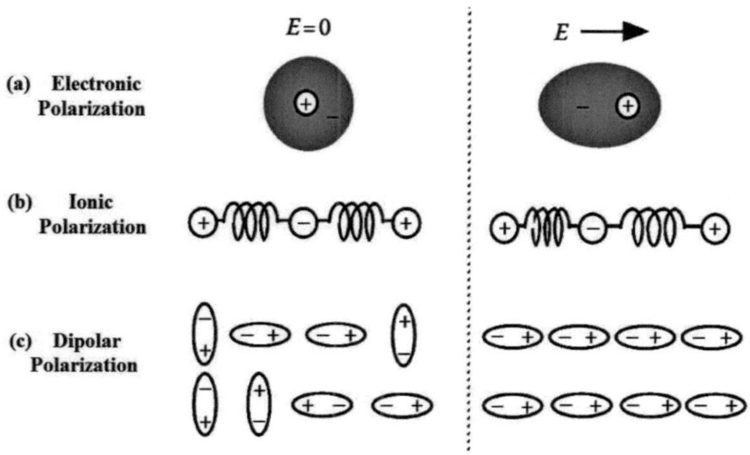

Figure 6.
Atomic contributions to electric polarization [18].

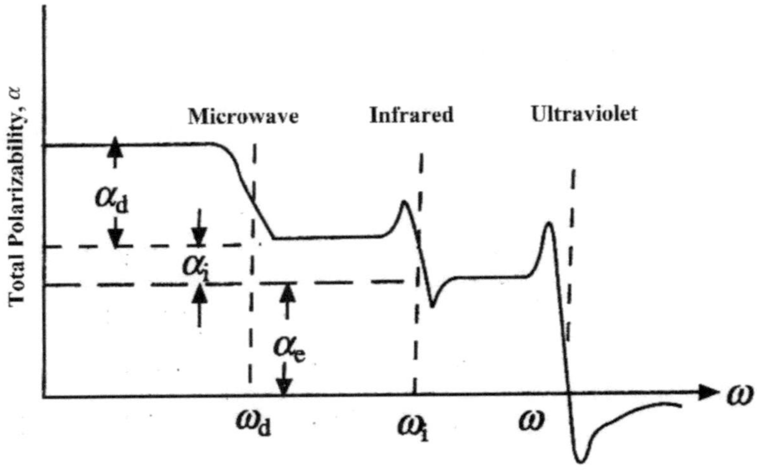

Figure 7.
Frequency dependence of the various contributions to the polarizability [6].

If the relative displacement of the positive and negative ions is d and the charge on each ion is q, then the dipole moment per molecule is $\bar{p} = qd$ and the ionic polarization becomes

$$P_{ion} = Nqd \tag{17}$$

where N is the number of atoms per unit volume.

iii. Dipolar polarizability (α_d): The dipolar polarizability, also called orientational polarizability, is important only in materials that contain complex ions having permanent dipole moment. In the absence of an external electric field, the dipoles have random orientations, and there is no net polarization. However, when the electric field is applied, the dipoles orient themselves along the direction of the field and produce dipolar or orientational polarization, as shown in **Figure 6c**. Such an orientation is opposed by the thermal agitation which tends. According to Debye's quantum theory, dipolar polarizability (α_d) per dipole is given by

$$\alpha_d = \frac{\bar{p}}{E} = \frac{p^2}{3kT} \tag{18}$$

where k is the Boltzmann's constant, T is the absolute temperature, and p is the dipole moment of the atom. The polarization contributed by electronic polarizability (α_e) and ionic polarizability (α_i) is called distortion polarization. Since α_e and α_i are temperature independent, the part of dielectric constant depending on them is essentially independent of the temperature. The contribution to the polarization made by dipolar polarizability (α_d) which is a function of temperature in accordance with Eq. (18). The contributions to the total polarizability (polarizability versus frequency curve) are shown in **Figure** 7 [2–17].

We find that in the optical frequency range, the dielectric constant (ε_0) arises entirely due to the electronic polarizability. The ionic and dipolar contributions are small at high frequencies because of the inertia of the ions and molecules.

5. Classification of ferroelectrics

Ferroelectric crystals have been classified into the following types [1–6].

1. According to the chemical composition of the crystal.

    a. Hydrogen bonded and its isomorphs such as KH_2PO_4 (KDP), triglycine sulphate (TGS), Rochelle salt (RS) and lead hydrogen phosphate (LHP) etc.

    b. Double oxides such as $BaTiO_3$, potassium niobite ($KNbO_3$) and lithium niobate ($LiNbO_3$) etc.

2. Based on the number of directions allowed to the spontaneous polarization are of two types.

    a. Single-axis of polarization such as Rochelle salt, KDP etc.

    b. Several-axes of polarization such as $BaTiO_3$ etc.

3. According to the existence or lack of centre of symmetry in non-polar phase.

    a. Non-centre of symmetrical non-polar phase such as KDP and Rochelle salt.

    b. Centre of symmetrical non-polar phase such as $BaTiO_3$ and TGS crystals, etc.

4. According to the nature of the phase change.

    a. Order–disorder type such as KDP, RS, TGS, LHP and CsH_2PO_4 (CDP) etc.

    b. Displacive type such as $BaTiO_3$, $LiNbO_3$ and $KNbO_3$ etc.

In the order–disorder group of ferroelectrics, the ferroelectric phase transition is associated with an individual ordering of ions. These are the crystals that contain H-bonds and in which the motion of protons is related to the ferroelectric properties. The examples are KH_2PO_4, RS, TGS, CsH_2PO_4, $PbHPO_4$ and RbH_2PO_4, etc. The displacive group of ferroelectrics is the one in which the ferroelectric phase transition is associated with the displacement of a whole sublattice of ions of one type relative to a sublattice of another type. The displacive type ferroelectrics possess perovskite ABO_3 type structures. Examples are $BaTiO_3$, $LiNbO_3$ and $KNbO_3$, etc. Consider the case of $BaTiO_3$ crystal, as shown in **Figure 8**. The unit cell is cubic with Ba^{2+} ions occupying at the corners, O^{2-} ions occupying the face centres and Ti^{4+} ion occupying the body centre of the cube. Thus, each Ti^{4+} ion is surrounded by six O^{2-} ions in an octahedral configuration. Above the Curie temperature ($T > T_c$), the prototype crystal structure is cubic, the centres of gravity of positive and negative charges exactly coincide with each other to produce a net dipole moment is zero. Below the Curie temperature ($T < T_c$), the structure is slightly deformed with Ti^{4+} at the body centre while Ba^{2+} ions at cube corners slightly move upwards, and the structure becomes tetragonal with centres of the positive (+) and negative (−) charges not coinciding with each other.

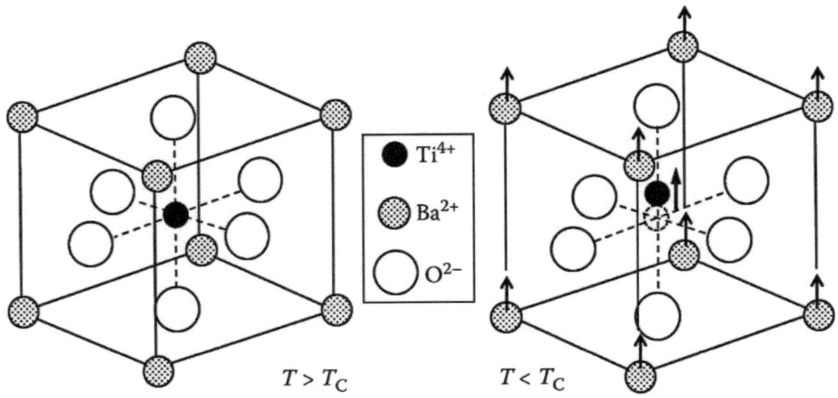

Figure 8.
Structures of $BaTiO_3$: $T > T_c$ left and $T < T_c$ right [18].

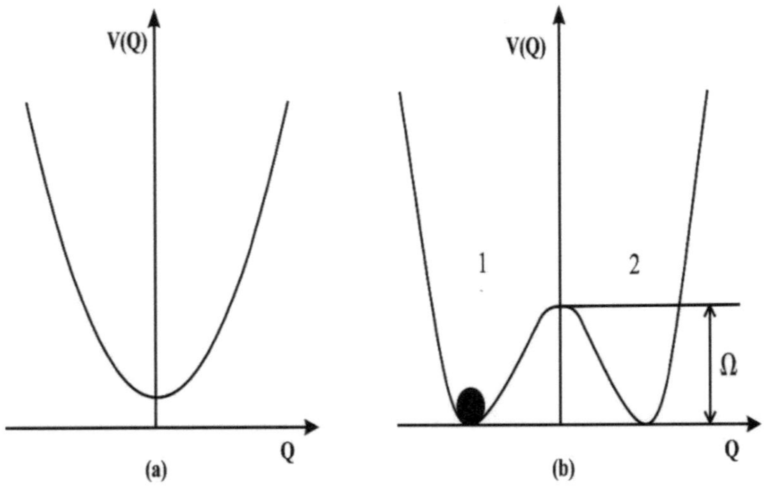

Figure 9.
Single cell potential for (a) displacive (b) order–disorder ferroelectrics [19].

This situation leads to net dipole moment and hence produces the spontaneous polarization (P_s) of the crystal [6–18]. **Figure 9** shows that in displacive ferroelectrics active atom has a single potential, while in the order–disorder ferroelectrics, the active atom has a double-well potential [19]. In the order–disorder systems, the proton can tunnel through the barrier, which separates the two minima of the potential energy in the hydrogen bond, and the ground state of the system splits into two levels separated by an energy Ω. The magnitude of Ω depends on the overlap of the wave functions appropriate to a proton located in each of the two separated minima.

6. Thermodynamics of ferroelectricity

Many of the experimental results on the macroscopic properties of ferroelectrics such as polarization and dielectric constant as well as their temperature, electric field and pressure dependence, etc. In 1920, Valasek [20] was discovered ferroelectricity in Rochelle salt. Thereafter, for about twenty years, the mechanism of ferroelectricity remained a mystery. In the period between 1935 and 1938, ferroelectricity in KDP crystal and its isostructural crystals was observed [21]. In order to explain ferroelectricity in KDP, a microscopic model Hamiltonian of disordered proton compositions was proposed, based on which a pseudospin model was developed by Slater [22] and Takagi [23] later independently. Ferroelectricity in Barium titanate ($BaTiO_3$) was reported in 1945–1946 [24, 25], and a microscopic model of Ti^{4+} ion displacement was proposed by Slater in 1950 [26]. Devonshire [27] developed a phenomenological approach based on Landu-Ginburg phase transition theories [28, 29] to explain ferroelectric phase transitions. Later on, the thermodynamic theories of piezoelectricity were summarized by Cady [30] and Mason [31] independently. The most important concept in the theory of solid-state phase transitions is the concept of a "soft mode", which was developed on the basis of lattice dynamics by Cochran [32, 33] and Anderson [34]. According to the concept of a "soft mode", ferroelectric order stems from the instability of a transverse vibrational mode or a ferroelectric mode. Detailed lattice dynamic calculations for ferroelectric crystals and more rigorous mathematical treatments of the soft mode in ferroelectrics and anti-ferroelectrics have been made by Blinc and Zeks [35] and others [36]. Later on, ten years ago, it was believed that there were two different types of ferroelectric phase transition mechanism: displacement type and order–disorder type. However, several ferroelectric phenomena discovered that could be explained neither by a displacement type mechanism nor by an order–disorder type mechanism unequivocally. Therefore, several unified models based on a combination of both mechanisms have been proposed [37], such as in the general model developed by Stamenkovic et al. [38], two basic ordering parameters associated with the motion of active atoms etc. Based on the theory of phase transition of Landau-Ginzburg [28, 29] and Devonshire [27], developed the phenomenological theory of ferroelectricity by choosing the polarization as an order parameter. The most convenient treatment of the ferroelectric phase transition by using the elastic Gibbs function G_1 as a state function of the ferroelectric system and the temperature (T), stress (Λ) and polarization (P) as independent variables [39, 40]. The Gibb's free energy function is expressed as

$$G_1 = U - T\sigma + \sum_{i,j} S_i \Lambda_j \tag{19}$$

where U is the internal energy of the system, T is the temperature and σ is the entropy, S_i and Λ_j are the ith and j^{th} component of mechanical strain and stress. Making the use of a differential form of the internal energy

$$dU = Td\sigma - \sum_{i,j} \Lambda_j dS_i + \sum_{n,m} E_n dP_m \tag{20}$$

where E_n is the components of the electric field. Therefore, we can write the differential form of elastic Gibbs function

$$dG_1 = -\sigma dT + \sum_{i,j} S_i d\Lambda_j + \sum_{n,m} E_n dP_m (i, j = 1, 2, \ldots 6; m, n = 1, 2, 3) \tag{21}$$

where E_n is the components of the electric field, we have $E_n = \left(\frac{dG_1}{dP_n}\right)_{T,\Lambda}$ and $S_i = -\left(\frac{dG_1}{d\Lambda_i}\right)_{T,P}$. Since the Gibbs free energy density $G = G_1 - E_n P_m$, the stable state of the system can be determined by the minimum of Gibb's free energy (G). If T and Λ_i are constants, G_1 is a function of the polarization P (if G_1 and P_n are known, then E_n are entirely determined).

6.1 Equation of state

We consider a ferroelectric crystal having an intrinsic spontaneous polarization (P_s) along a specific-axis in the space coordinate system and assume that the external pressure is constant (say one atmosphere). As G_1 of the system is not changed by reversing the direction of the axes of space coordinate system, G_1 is independent of the direction of polarization (P). Thus, G_1 is an even function of P. Therefore, we can expand G_1 as a power series, in even powers of polarization P and neglecting the odd powers of P for symmetry reasons

$$G_1(T,P) = G_{10}(T) + \frac{1}{2}\beta(T)P^2 + \frac{1}{4}\xi_1(T)P^4 + \frac{1}{6}\xi_2(T)P^6 + \dots \quad (22)$$

where $G_{10}(T)$ is the value of elastic Gibb's free energy (G_1) of the system at $P = 0$ and in general, the coefficients G_{10}, β, ξ_1 and ξ_2 ... ., are the functions of temperature (T). A stable state of a thermodynamic system is characterized by a minimum value of the Gibbs free energy G. We have $G_1 = G$ when $E = 0$, G can be replaced by G_1. When a crystal exhibits a stable spontaneous polarization (P_s) at a certain temperature, the conditions for a minimum of G_1 are

$$\left(\frac{\partial G_1}{\partial P}\right)_{P_s} = 0, \left(\frac{\partial^2 G_1}{\partial P^2}\right)_{P_s} > 0 \; or \; \left(\frac{\partial E}{\partial P}\right)_{P_s} = \chi^{-1} > 0 \quad (23)$$

Using Eq. (22) into (23), we obtained the equation of state for the ferroelectric system of the form

$$P_s(\beta + \xi_1 P_s^{\,2} + \xi_2 P_s^{\,4}) = 0 \quad (24)$$

$$\chi^{-1} = (\beta + 3\xi_1 P_s^{\,2} + 5\xi_2 P_s^{\,4}) > 0 \quad (25)$$

Eq. (24) has two roots: (i) the first root $P_s = 0$ corresponds to a paraelectric phase and (ii) the second root $P_s \neq 0$ corresponds to a ferroelectric phase.

6.2 Paraelectric phase

Suppose spontaneous polarization $P_s = 0$; from Eq. (25), the reciprocal of the dielectric susceptibility can be explained as.

$$\chi^{-1} = \beta(T) > 0 \quad (26)$$

It is obvious that the value of β have a positive value when a stable state of the crystal is a paraelectric phase. Therefore, the boundary conditions at the critical temperature is $(\beta(T))_{T_0} \geq 0$. Expanding $\beta(T)$ as a Taylor's series in $(T - T_0)$ and taking into account only the first-order term in $(T - T_0)$, we have

$$\beta(T) = \frac{(T - T_0)}{C} \quad (27)$$

By combining Eqs. (26) and (27) we get

$$\chi = \frac{C}{(T - T_0)} \tag{28}$$

where C is the Curie–Weiss constant. This is the Curie–Weiss law, which applies to the dielectric susceptibility in a paraelectric phase. In the case of spontaneous polarization $P_s \neq 0$, we will see that the one result corresponds to a second-order phase transition when $\xi_1 > 0$, and that the other result to a first-order phase transition when $\xi_1 < 0$.

6.3 Second-order phase transitions

Consider first case $\xi_1 > 0$. The roots of Eq. (24) for $P_s^2 \neq 0$ are

$$P_s^2 = \frac{\left[-\xi_1 \pm \left(\xi_1^2 - 4\beta\xi_2\right)^{1/2}\right]}{2\xi_2} (\xi_1, \xi_2 > 0, \beta < 0) \tag{29}$$

One of the two roots of P_s^2 is always negative and corresponding to an imaginary value of P_s. When $\beta < 0$, we may get a positive root of P_s^2 and corresponding to a real value of P_s. However, for a real ferroelectric crystal $|\beta|\xi_2 \ll \xi_1^2$.

6.4 Susceptibility

When the temperature is below T_c. From Eq. (26), we get reciprocal of the dielectric susceptibility

$$\chi^{-1} = \left(\frac{\partial E}{\partial P}\right)_{P=P_s} = \beta + 3\xi_1 P_s^2 + 5\xi_2 P_s^4 \tag{30}$$

The term of $\xi_2 P_s^4$ can be neglected when the temperature is below and near T_c because then P_s is very small and using the relation $P_s^2 = -\frac{\beta}{\xi_1}, (\beta < 0, \xi_1 > 0\ )$ into Eq. (30) we have.

$$\chi^{-1} = -2\beta = -2\frac{(T - T_c)}{C}, (T < T_c) \tag{31}$$

We plot the reciprocal of the susceptibility as a function of temperature in **Figure 10a**. This theoretical plot agrees well with the experimental data; the slope of the χ^{-1} curve in the ferroelectric phase is twice that of the χ^{-1} curve in the paraelectric phase.

6.5 Free energy

For $\xi_1 > 0$, using Eq. (22), $G_1 - G_{10}$ is plotted as a function of the polarization (P) at several temperatures ($T_{c1} < T_c < T_{c2}$) in **Figure 10b**. Since the sign of β, positive at $T = T_c$ and at T_{c2}, turns negative at T_{c1}, the curve representing ($G_1 - G_{10}$) changes from a minimum at T_{c2} to maximum at T_{c1} at $P = 0$. At T_{c1}, the two minima of the free energy ($P \neq 0$) corresponds to stable ferroelectric states.

6.6 Ferroelectric hysteresis loop

Using the value of $\beta = \frac{(T-T_c)}{C}$ and $E_i = \left(\frac{\partial G_1}{\partial P_i}\right)_{T,\Lambda}$ impiles

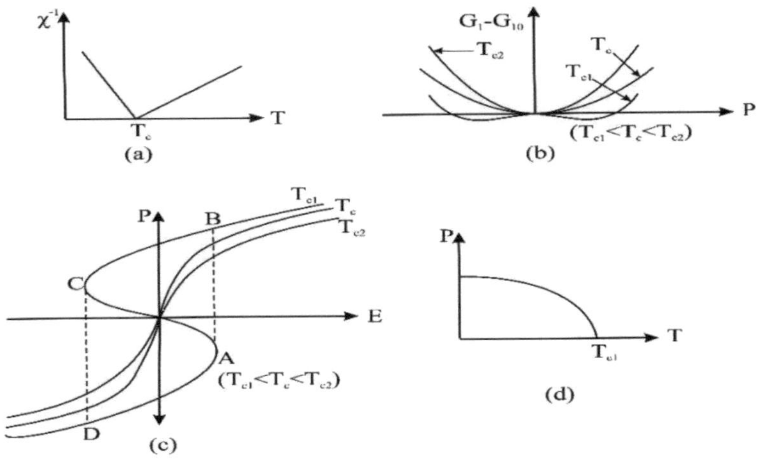

Figure 10.
Functional relations of (a) χ^{-1} *versus* T; *(b)* $G_1 - G_{10}$ *versus* P; *(c)* P *versus* E *and (d)* P *versus* T *near the second-order phase transition [40].*

$$E = \beta P + \xi_1 P^3 + \xi_2 P^5 \tag{32}$$

We get $(P - E)$ curve at different temperatures corresponding to different β values are displayed in **Figure 10c**. In fact, at temperature $T_{c1}(<T_c)$, the segment in the curve from point A to point C corresponds to an unstable state, since the slope in this segment corresponds to $\beta > 0$. The experimental curves always jump directly from the state A to the state B and also directly from C to D. Thus, the observed results are hysteresis loop.

6.7 Spontaneous polarization

When $E = 0$, we putting Eq. (27) into relation $P_s^2 = -\frac{\beta}{\xi_1}$, $(\beta < 0, \xi_1 > 0)$, we get

$$P_s^2 = \frac{(T_c - T)}{\xi_1 C} \tag{33}$$

The function P_s changes continuously with temperature and becomes zero at T_c as shown in **Figure 10d**. This theoretical curve agrees with the experimental results in ferroelectric crystals exhibiting second-order phase transitions.

6.8 First order phase transitions ($\xi_1 < 0$)

As explained above the condition for the occurrence of spontaneous polarization (P_s) is that β should be negative while ξ_1 should be positive and there is second-order transition, for first-order transition, the coefficient β is negative and also ξ_1 is negative as temperature is lowered. The Gibbs free energy curves with function of polarization (P) at different temperatures for this transition are shown in **Figure 11a**. It is obvious that the polarization state $(P \neq 0)$ is stable at the temperature $T_{c1}(<T_c)$ in the **Figure 11b**. The P-E curves at various temperatures are plotted in **Figure 11b** from the hysteresis loop occurs at $T_{c1}(<T_c)$. For $E = 0$, the spontaneous polarization (P_s) satisfies the Eq. (28) and from the Eq. (22), using the condition $G_1 = G_{10}$ we get the following relations given by.

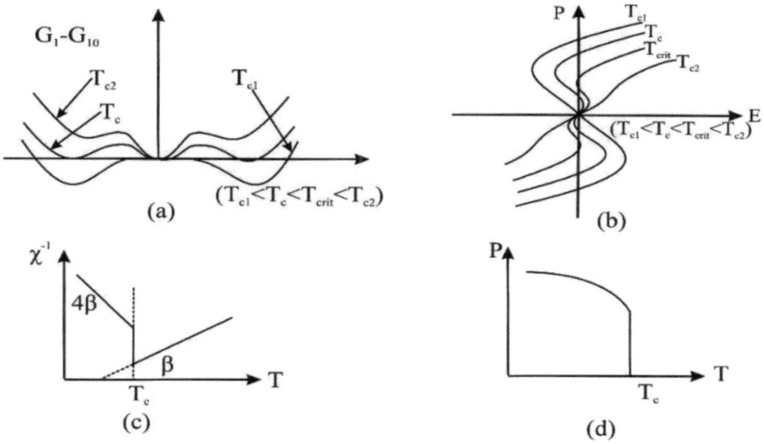

Figure 11.
Functional relations of (a) $G_1 - G_{10}$ versus P; (b) P versus E; (c) χ^{-1} versus T and (d) P versus T near the first-order phase transition [40].

$$P_s^2(T_c) = \frac{3|\xi_1|}{4\xi_2}, \beta = \frac{3\xi_1{}^2}{16\xi_2}, P_s^4(T_c) = \frac{3\beta}{\xi_2} \tag{34}$$

Combing the Eq. (30) with Eq. (34) and keeping in the mind that $\xi_1 < 0$ in the second term of Eq. (30) we get.

$$\chi^{-1} = 4\beta = 4\frac{(T - T_c)}{C} \tag{35}$$

At the transition temperature (T_c), the χ^{-1} in a first order transition is not zero but it is positive quantity as can be seen from Eq. (35). The variation of temperature dependence of inverse (χ^{-1}) above and below the transition temperature (T_c) is shown in **Figure 11c**. The χ^{-1} at a temperature just below T_c is four times that at a temperature above T_c. The curve for the polarization (P) with temperature (T) is shown in **Figure 11d**.

7. Model theories of ferroelectricity

Various physicists have developed model theories of ferroelectricity. An introductory idea of model theories that have been developed to explain the phenomenon of ferroelectricity is given below. Many experimental and theoretical attempts were made to explain the phenomenon of ferroelectricity in single and polycrystals and proposed a number of theories. The first theoretical explanation of the ferroelectric properties of Rochelle salt was proposed by Kurchatov [41]. Slater [22] put forward the first molecular theory of ferroelectricity and suggested that the ferroelectric behaviour in KDP and Rochelle salt is principally due to the ordering of H-bonds. A general theory of ferroelectricity established at that time by Cochran's lattice dynamic theory [33] and Lines statistical theory [42] have provided for major understanding of the ferroelectric phenomena of ferroelectricity.

A simple order–disorder model Hamiltonian was proposed by Mason [43]. In this model, a proton motion along an H-bond. A proton may transfer from one-well to another, and vice versa, stochastically over a potential barrier, and this model does not explain the isotope effect. Blinc [44] introduced the concept of proton tunneling motion between the two equilibrium sites in the double-well minimum O-H–O bond potential. The Blinc's [44] concept was quickly put into the simple formalism of the pseudospin-model by De Gennes [45] and independently by Matsubara [46]. De Gennes [45] showed that the proton in a double minimum type potential has described by a one-half pseudospin. The pseudospin model Hamiltonian developed to explain the proton system and the triggering of phase transition can be ascribed to the ordering of the proton in the double-well potential. The dipoles formed by the proton ordering make a small contribution to the spontaneous polarization (P_s) which is the result of heavy atoms displacements along the ferroelectric axis. The major contribution to the spontaneous polarization (P_s) is the displacement of heavy atoms projected on c-axis. To account for the displacement of the heavy atoms, Kobayashi [47] included pseudospin lattice interaction into the pseudospin model. The net result of this approach was to enhance the effective dipolar proton-proton interaction.

Later on, Arefev et al. [48] and Brout et al. [49] suggested that essentially the same concept could also be applied such that KDP etc. In this case, where the permanent electric dipoles move-in a potential with more than one equilibrium position and the soft mode collective excitations are not phonons but rather the unstable pseudospin phonon coupled-wave. Kaminow [50] experimentally confirmed the existence of soft mode in KDP crystal, and the other investigators also confirmed the soft mode in other ferroelectric crystals. It is now well recognized that the several interesting properties of ferroelectrics are associated with the high-temperature dependence of the soft mode. Cowely [51] has given a microscopic theory of ferroelectricity in which the temperature dependence of the normal mode (soft mode) arises from anharmonic interactions between normal modes. These anharmonic interactions in a crystal are quite small, at least at low temperatures, so that an anharmonic crystal provides an example of the many-body system in which interactions between the elementary excitations are both small and non-singular. In the harmonic approximation, the equations of motion of the shell model can be obtained from a quadratic function of the displacements of the ions and of the electronic dipoles produced on the ions during the lattice vibrations. The anharmonic interactions arise from the cubic and higher terms in potential function, and in general, there will be anharmonic interactions between all the displacements and all the dipoles. In the ferroelectrics, the root at wave vector $\vec{k} = 0$ is imaginary in the harmonic approximation showing the instability of the lattice [51]. This indicates that the harmonic forces alone are not sufficient to stabilize the system at any temperature. The stabilization of the mode can only be brought about by a consideration of anharmonic interactions terms. The anharmonic interactions thus play a fundamental role as regards the stability of the crystal system. A number of physical properties of solids could be well explained by considering the effects of phonon anharmonic interactions.

A very successful attempt has been made to give a microscopic theory of ferroelectric crystals given by Cochran [33, 52]. This lattice dynamical theory is based on the hypothesis that the ferroelectric transitions are the results of the instability of the crystal lattice with respect to one of the homogeneous (wave number $k = \frac{2\pi}{\lambda} = 0$) transverse optic mode. If a crystal is fully or partly ionic, lattice vibrations are accompanied by polarization oscillations having an equal frequency which provide a Lorentz field called local field interacting with the ions through long-range Coulomb forces. The crystal becomes unstable for one particular mode of vibration at which the long-range forces are equal and opposite to the short-range forces. A relation given by Lyddane et al. [53] explains the relation between the ferroelectric properties and the thermodynamic properties of the crystals. For $k = 0$ modes of diatomic crystal, Lyddane et al. [53] relation gives the ratio of the static dielectric constant (ε_s) of the crystal to the high-frequency dielectric constant (ε_e) in terms of the frequencies of the longitudinal optical (ω_L) and transverse optical (ω_T) of infinite wavelength

$$\frac{\varepsilon_s}{\varepsilon_e} = \frac{\omega_L{}^2}{\omega_T{}^2} \tag{36}$$

When $\omega_T{}^2 = 0$, we get $\varepsilon_s = \infty$. Cochran [32] developed a more general case in which there are n atoms in the elementary cell as

$$\frac{\varepsilon_s}{\varepsilon_e} = \prod_j^n \frac{(\omega_j{}^2)_L}{(\omega_j{}^2)_T} \tag{37}$$

This Eq. (36) produces one essential anomaly needed to explain a ferroelectric transition. In order to complete understanding the ferroelectric behaviour, it is necessary to investigate the temperature of $\omega_T{}^2$. In the ferroelectric crystal, the static dielectric constant (ε_s) obeys Curie–Weiss law above $T_c(T > T_c)$. Eq. (37) was derived for harmonic forces. In order to derive the temperature dependence of $\omega_T{}^2$ it will be necessary to introduce anharmonic interaction, which shows little effect on any mode other than$k = 0$ mode, and only that mode exhibits any anomalous behaviour. Thus Eq. (37) through (36) implies that the transverse optical mode $\omega_T{}^2$ have anomalous temperature dependence given by the Curie–Weiss law by a relation $\omega_T{}^2 = K(T - T_c)$, where the coefficient K is constant related to the temperature dependence dielectric constant $\varepsilon(T)$ through the Lyddane et al. [53] relation.

8. Applications of ferroelectrics

Ferroelectric materials have been extensive applications [3, 54] in a large number of areas due to their peculiar and interesting properties such as high permittivity capacitors ($BaTiO_3$), ferroelectric non-volatile FeRAM memories (due to bi-stable polarization in modulation and deflector), pyroelectric sensors, piezoelectric and electrostrictive transducers (TGS crystal), electrooptic and optoelectronic devices (due to their non-linear polarizability), thermistors, storage and laser devices, sensors, resonators and actuators which have revolutionized consumer electronics, automobile industry, biomedical diagnosis, underwater acoustic technology, defense-related sectors, gas sensing devices and surface acoustic wave technology, etc. The major areas of applications [3, 54] of ferroelectrics have received a great deal of attention amongst all the above capacitors, ferroelectric memories, pyroelectric sensors, piezoelectric, electrostrictive transducers, electrooptic devices and thermistors. The basic specifications required for capacitors are small size, large capacitance (materials with a large dielectric constant are desired). High frequency characteristics (ferroelectrics with a high dielectric constant are sometimes associated with dielectric dispersions, which must be taken into account for practical applications). Ferroelectric relaxors such as $Pb(Mg_{1/3}Nb_{2/3})O_3$ and $Pb(Zn_{1/3}Nb_{2/3})O_3$ are some examples of these applications.

The bi-stable polarization of ferroelectrics makes them useful for binary memory systems. There are volatile and non-volatile memory devices in erasable semiconductor memories. Non-volatile memory does not require a holding voltage. Dynamic random-access memory (DRAM), which is widely used because of its high integration capability, is an example of volatile memory. Data stored in these memories are lost when the electric power is shut off. To record information of polarization may be reversed or reoriented by application of an electric field greater than the coercive field. For erasure, the polarization can be returned to its original state with an applied field of opposite polarity. To read the stored information, it is retrieved by electrical or optical means. Optical memory is an electrically addressed light valve. For example, $BaTiO_3$, $(Pb,La)(Zr,Ti)O_3$ and $Pb_5Ge_3O_{11}$ single crystals are extensively used as light values. When a ferroelectric thin film with a large polarization-electric field hysteresis is used as the memory capacitor, non-volatile memory is realized. When a voltage is applied to the gate and the field-effect transistor (FET) assumes the "on" state, a pulse voltage to the drain generates a drain current dependent on the remanent polarization state. A large electric field is applied to a ferroelectric film in every process in the ferroelectric RAM (FeRAM), the polarization hysteresis characteristic degrades with increasing cycles. This problem of ferroelectric films needs to be overcome for non-volatile memory applications. The development of the ferroelectric memory started with DRAMs is composed of a FET and memory capacitor, then moved into FeRAMs and is now focused on metal ferroelectric semiconductor field-effect transistors (MFSFETs). $BaTiO_3$, $LiNbO_3$ and KH_2PO_4 crystals, etc., are some examples of these applications.

The pyroelectric properties of polar materials were studied a long time ago, and such materials were belonged as electric stones, measuring the current or voltage response of a crystal to a temperature change, either by continuous heating or by the absorption of sinusoidally modulated radiation. This is basically due to the temperature dependence of the spontaneous polarization of a polar material. The pyroelectric sensors are widely used for monitoring temperature or infrared radiation (IR). Practical applications of the pyroelectric material in temperature sensors and infrared (IR) light detectors lead to some commercial making of ferroelectric ceramics. Pyroelectric detectors can be used to record infrared images. The converse effect is called the electrocaloric effect, which may be a future cooling system. Materials such as TGS, $LiTaO_3$, $Sr_{1/2}Ba_{1/2}Nb_2O_6$ etc. are some examples. Another important application of piezoelectric devices. Certain materials produce electric charges on their surfaces when mechanical stress is applied. The induced charges are proportional to the mechanical stress. This is called the piezoelectric effect, which was discovered in Quartz by Pierre and Jacques Curie in 1880. Materials showing this phenomenon also conversely have a geometric strain proportional to an applied electric field showing the converse piezoelectric effect, discovered by Gabriel Lippmann in 1881. The root of the word "piezo" means "pressure" in Greek. Hence the original meaning of the word piezoelectricity implied "pressure electricity". The phenomenon of Piezoelectricity is widely utilized in the fabrication of various devices such as sensors, transducers, actuators, surface acoustic wave (SAW) devices, frequency control devices etc. Quartz, $BaTiO_3$, $(Pb, Sm)TiO_3$, $LiNbO_3$, and $LiTaO_3$ etc., are some materials that can be used for these applications.

Acknowledgements

Authors are thankful to Prof. S.C. Bhatt, Prof. P.D. Semalty for valuable suggestions and discussions.

Author details

Muzaffar Iqbal Khan* and Trilok Chandra Upadhyay
Department of Physics, Hemvati Nandan Bahuguna Garhwal University (A Central University), Srinagar (Garhwal), Uttarakhand, India

*Address all correspondence to: muzaffariqbalkhan786@gmail.com

References

[1] R.N.P Choudhary and S.K. Patri, *Dielectric Materials: Introduction, Research and Applications*, (Nova Science Publishers, 2009).

[2] F. Jona and G. Shirane, *Ferroelectric Crystals*, (Pergamon Press, London, 1962).

[3] M.E. Lines and A.M. Glass, *Principles and Applications of Ferroelectrics and Related Materials*, (Clarendon Press, Oxford, 1977).

[4] H.D. Megaw, *Ferroelectricity in Crystals*, (Methuen Publishing Co. Ltd., London 1957).

[5] E.T. Jaynes, *Ferroelectricity*, (Princeton University Press, Princeton, 1953).

[6] M.A. Wahab, *Solid State Physics*, (Narosa Publishing House, 2005).

[7] Richard J.D. Tilley, *Understanding Solids*, (Wiley & Sons, Second Edition, 2013).

[8] A.R. West, *Basic Solid-State Chemistry*, (John Wiley & Sons, 1999).

[9] E. Fatuzzo and W.J. Merz, *Ferroelectricity*, (North-Holland Pub., Amsterdam, 1967).

[10] S.J. Blundell and M.B. Katherine, *Concepts of Thermal Physics*, (Oxford University Press, 2008).

[11] J. Grindlay, *Theory of Ferroelectricity*, (Pergemon Press, London, 1967).

[12] T. Mitsui, *An Intro. to Phy. of Ferroelectrics*, (Gordon & Breach Pub., New York, 1976).

[13] A.K. Bain and P. Chand, *Ferroelectric: Principles and Applications*, (Wiley-VCH, Germany, 2017).

[14] J.C. Burfoot, *Ferroelectrics: An Introduction to Physical Principles*, (Van Nostrand & Co. Ltd. Princeton, 1967).

[15] P.W. Forsberg, *Piezoelectricity, Electrostriction and Ferroelectricity*, (Springer Verlag, Berlin, 1957).

[16] A. J. Dekkar, *Solid State Physics*, (Palgrave, London, 1952).

[17] W. Kanzing, *Ferroelectrics and Anti-ferroelectrics in Sol. St. Phy.*, (Acad. Press, 1957).

[18] K. Uchino, *Ferroelectric Devices*, (CRC Press, Taylor & Franics Group, 2010).

[19] R. Blinc & B. Zeks, *Advances in Physics*, **21(93)**, 693–757 (1972).

[20] J. Valasek, *Physical Review*, **17(4)**, 475 (1921).

[21] G. Busch and P. Scherrer, *Naturwissenschaften*, **23(43)**, 737 (1935).

[22] J.C. Slater, *Journal of Chemical Physics*, **9(1)**, 16–33 (1941).

[23] Y. Takagi, *Journal of Physical Society of Japan*, **3(4)**, 271–272 (1948).

[24] A. Von Hippel, *Reviews of Modern Physics*, **22(3)**, 221 (1950).

[25] B. Wul and J.M. Goldman, *CR Acad. Sci. URSS*, **51**, 21 (1946).

[26] J.C. Slater, *Physical Review*, **78(6)**, 748 (1950).

[27] A.F. Devonshire, *Advances in Physics*, **3(10)**, 85–130 (1954).

[28] L.D. Landau, *Z. Sowjetunion*, **11**, 545 (1937).

[29] V.L. Ginzburg, *Zh. Eksp. Teor. Fiz*, **19**, 36–41 (1949).

[30] W.G. Cady, *Piezoelectricity*, (McGraw-Hill, New York, 1946).

[31] W.P. Mason, *Piezoelectric Crystals and Their Applications to Ultrasonics*, (D. Van Nostrand Co., New York, 1950).

[32] W. Cochran, *Physical Review Letters*, **3(9)**, 412 (1959).

[33] W. Cochran, *Advances in Physics*, **9 (36)**, 387–423 (1960).

[34] P.W. Anderson, *Fizika Dielektrikov*, (Akad. Nauk SSSR, Moscow, 1960).

[35] R. Blinc and B. Zeks, *Soft Modes in Ferroelectrics and Antiferroelectrics*, (Elsevier, New York, 1974).

[36] K.R. Rao and S.L. Chaplot, *Ferroelectrics*, **29(1)**, 7–10 (1980).

[37] S. Aubry, *Ferroelectrics*, **16(1)**, 313–316 (1977).

[38] S. Stamenkovic, et al., *Physical Review B*, **14 (11)**, 5080 (1976).

[39] T. Mitsui, I. Tatsuzaki and E. Nakamura, *An Introduction to the Physics of Ferroelectrics*, (Gordon and Breach, 1976).

[40] Xu. Yuhuan, *Ferroelectric Materials and Their Applications*, (Elsevier Sci. Pub., 1991).

[41] I.V. Kurchatov, *Seignette Electricity*, (Moscow, 1933).

[42] M.E. Lines, *Physical Review*, **177**, 797–819 (1969).

[43] W.P. Mason, *Physical Review*, **72(9)**, 854 (1947).

[44] R. Blinc, *Journal of Physics and Chemistry of Solids*, **13(3–4)**, 204–211 (1960).

[45] P.G. De Gennes, *Solid State Communications*, **1(6)**, 132–137 (1963).

[46] T. Matsubara, *Report of Res. Inst. for Solid State Physics*, Japan (1962).

[47] K. K. Kobayashi, *Journal of Physical Society of Japan*, **24(3)**, 497–508 (1968).

[48] I.M. Arefev, P.A. Bazhulin and T.V. Mikhaltseva, *Sov. Phys. Sol. St.*, **7**, 1948 (1966).

[49] R. Brout, K.A. Muller and H. Thomas, *Solid State Communications*, **4**, 507 (1966).

[50] I.P. Kaminow, *Physical Review*, **138 (5A)**, A1539 (1965).

[51] R.A. Cowley, *Physical Review Letters*, **9(4)**, 159 (1962).

[52] W. Cochran, *Advances in Physics*, **10 (40)**, 401–420 (1961).

[53] R.H. Lyddane, R.G. Sachs and E. Teller, *Physical Review*, **59(8)**, 673 (1941).

[54] K. Uchino, *Ferroelectric Devices*, (CRC Press, 2009).

Chapter 3

Perovskite Ferroelectric

Paramjit Kour and Sudipta Kishore Pradhan

Abstract

The spectrums of properties exhibited by ferroelectric materials are dielectric, ferroelectric, piezoelectric and pyroelectric effect. This is the makes these materials to have a wide range of useful application. Infrared detectors are used pyroelectric effect of ferroelectric materials. It is used in nonvolatile memories due to have ferroelectric hysteresis. Its piezoelectric properties make them useful for actuator, radio frequency filter, sensor, and transducer. Ferroelectric capacitors are used, their good dielectric behavior. According to the necessity of the system they are available in different form such as single crystals, ceramics, thin film, and polymer, composite. The diversity of properties ferroelectric materials always attracted the attention of engineers and researchers. Size reduction of this material from micro to nanoscale established an enormous consideration to develop nanotechnology. Its vast use of different filed imposed the in detail research in adding to the development of processing and characterization method. This chapter will put some light on some fundamental principle of ferroelectricity, the list of perovskite materials and their application.

Keywords: PZT, $BaTiO_3$, $CaTiO_3$, PT, PLZT, Kb

1. Introduction

Like ferromagnetic materials, the functional properties of ferroelectric materials find wide range of applications, ranging from actuators and sensors to memory or optical devices. A ferroelctric class of materials cannot define in a single sentence. So before we define the ferroelectric materials, we should classify dielectric materials. Dielectric is belonging to a class of insulating materials that on the application of an electric field shows dielectric polarization [1]. Here the center of symmetry plays a significant role for their properties. Crystal structure with a center of symmetry have such an arrangement of atoms around a point or center that by the inversion, we can get the same arrangement of atoms in the crystal. Dielectric materials belong to a group of non Centro symmetric crystal structure. In 432 point group the entire non Centro symmetric point group shows piezoelectric properties [2]. The properties due to which voltage obtains form charge a separation in the face due to the mechanical stress and vice versa. Both direct and inverse piezoelectric effects have a wide range of application in electronic devices [3]. Barium titanate is an example of non-centrosymmetric piezoelectric material used in microphone and transducer [2–4]. In non centrosymmetric crystals there is an axis of symmetry, called polarity. These piezoelectric polar crystals are shown pyroelectricity. With changes in temperature there is a charge separation. The cells of polar structure have efficient dielectric polarization, so often called a spontaneous polarization. Either by stress or by a change in temperature the dipole moment of these polar structures is change. The is a charge separation in the surface, results in the spontaneous polarization. So the polar dielectric spontaneous polarization direction and magnitude can be modified with the applied stress. Zinc oxide, which belongs to family of polar dielectric shows wurtzitecrystal structure [5]. In this structure, between hexagonally packed oxygen ions layer, Zn2+ions are at the tetrahedral site.

Between these layers dielectric, whose spontaneous polarization depends on the direction and magnitude of the applied stress. Adequate amount of stress can change the direction of spontaneous polarization. But the exclusion of the stress does not bring back the original magnitude and direction of the spontaneous polarization. These families of polar dielectric are called ferroelectric. **Figure 1** shows the way the Centro symmetric, acentric, polar and ferroelectric, dielectrics are related to each other. Many review articles shown the history of ferroelectric [6–12]. Many great scientist open the path of discovery of the ferroelectric.

Pyroelctricity was studied by Brewster. Piezoelectricity was discovered by J. P. Curie. Debye, Boltzmann, Pockelsetc helps in conceptualizing the polarization in the dielectric. It was E. Schoridgener, who coined the term ferroelectricity, but JoesphValasek known for the discovery of ferroelectric. In 1920 the Rochelle salt (sodium potassium tartrate) shows spontaneous polarization which can be switch with the magnitude and direction of the applied field. This is the first manifestation of the ferroelectricity in a crystal. That is the trademark of ferroelectricity. Logically effectively the term ferroelectricity is defined as the switchable polarization between two or more stable state by the application of electric field. There exist some exceptions. Some semiconductor materials show ferroelectric properties. They do not posses electric polarization. In some ferroelectrics materials the spontaneous polarization cannot be switched with the electric field. This is either due to they are too conducting or reach the electrical break down first. This ferroelectric property first observed in Rochelle salt. But later it is observed in oxides, polymer, ceramics, and liquid crystal. When it is about ferroelectric property, the perovskite structure materials have a special importance. So in this chapter, we will discuss some perovskite ferroelectrics which are used in various electronics devices. According to the structure, there are five types of structure. (i) Organicpolymer (ii) Charge ordered ferroelectrics (iii) Magnetic order ferroelectric (iv) Corner-sharing oxygen octahedral (v) Hydrogen bonded radical compound (v) Ceramic polymer composites [11]. Among these ferroelectricgroup a mostly used ferroelectrics are the corner sharing oxygen octahedral oxides.

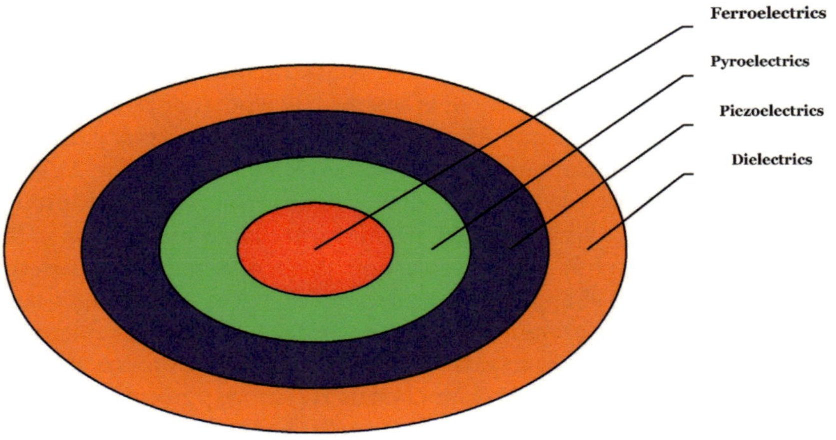

Figure 1.
Ven diagram of ferroelectric fit into different materials.

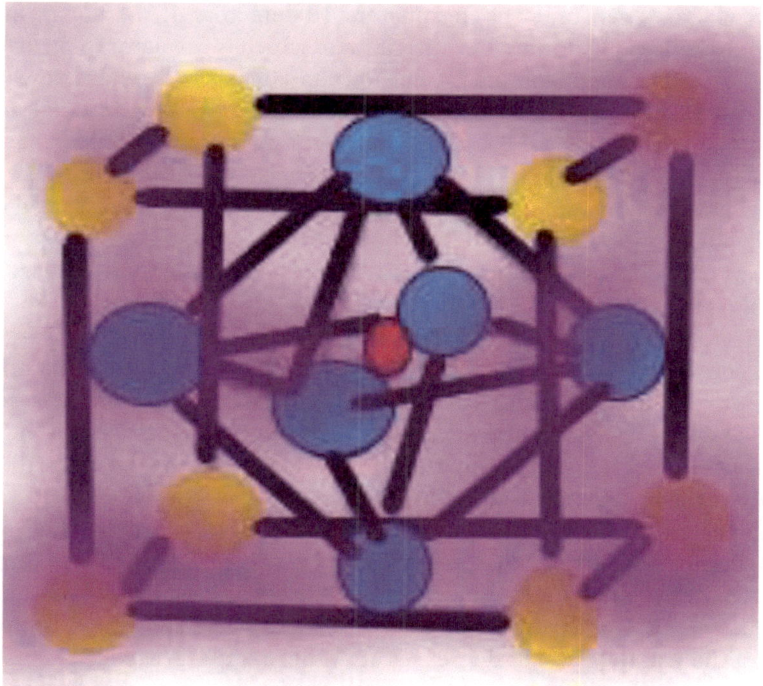

Figure 2.
Shows the schematic diagram of perovskite structure.

Basically, the corner sharing oxygen octahedral oxide structures is represented as $A^{a+}B^{b+}O_3$ [12]. **Figure 2** shows the schematic figure of O^{2-} ions corner sharing oxygen octahedral. The B^{b+} cation is seated within every octahedron. The b of the cation has the value lies in between 3 to 6. A^{a+} ions lie in the gap among the octahedral with its value of *a* is in between 1 to 3. A nonpolar lattice has been observed in prototype forms by the overlap of A^{a+}, B^{b+}, and O^{2-} ions geometric center. The total polarity of the lattice is obtained by displacement of A and B ions with respect to the O^{2-} ions. Due to change in temperature, phase transition takes place. This will result in displacement of ion results in a change in lattice structure. Spontaneous polarization will produce due to the displacement of ions in the arrangement of dipoles if there is no recompense pattern in the diploe.

2. Perovskite ferroelectrics

It is a family of a subgroup of corner sharing oxygen octahedral material's exhibiting ABO_3 structure. This family contains some mostly used piezoelectric and ferroelectric ceramics. Some member which is used in different field of Technology are strontium titanate (STO), barium titanate ($BaTiO_3$), ($SrTiO_3$), lead titanate ($PbTiO_3$), barium strontium titanate (BST), PZT, potassium niobate (KN) ($KNbO_3$) etc. Some of these perovskite are discussed in detail as follows.

2.1 Barium titanate ceramics

Barium titanate which a member of perovskite family exhibit good piezoelectric, ferroelectric and high dielectric constant. This FE ceramics are used in first piezoelectric transducer. But now it is mostly used in multilayer capacitors (MLCs) due to having a high dielectric constant. It is also used in positive temperature coefficient

Resistance (PTCR), sensor, PTC Thermistors, IR detectors, RAM, sonars and in electro optic devices. By doping Barium titanate which is insulator in pure form can be changed into a semiconductor. Barium titanate basically white powder of inorganic compound. With a decrease in temperature the octahedral TiO_6 undergoes distortion gives rise to five structural phase transition from hexagonal, cubic, tetragonal, orthorhombic, and rhombohedra. Very large spontaneous polarization and high dielectric constant are observed in this phase due to the distorted octahedral. At Curie point, i.e. above 120°C the distorted octahedral of TiO_6 comes to equilibrium result in an isotropic cubic structure [13, 14] so only this phase does not exhibit FE properties.

Capacitor, MLC etc. used first $BaTiO_3$ as the dielectric ceramic with large dielectric constant and dielectric loss for manufacturing. The factor that affects the dielectric properties of the materials is not only its structure but also its synthesis route which will reflect in its size of the grain, density, purity etc. [15]. The Applied frequency of the electric field, temperature and do pants also affects the dielectric properties of the materials [16–23].

The dielectric constant of Barium titanate prepared by any method increase depends on the grain size and distribution of grains [17, 24]. At room temperature the frequency dependent dielectric constant decreases at low frequency and then increases slightly and become constant at high frequency. Temperature dependent dielectric constant decreases at higher temperature [25]. Ions of different size can be can be added to the perovskite structure. So doping at both A and B site of this perovskite structure is used to tolerate its electrical properties [26]. At A site mono, die and trivalent acceptor do pants are substituted to produce P-type semiconductor, whereas N type semiconductor are obtained by donor dopant of tri, tetra and pentavalent ions at the B site of the perovskite. The concentration of the dopants also affects the electrical properties of Barium titanate. Increase in concentration of Donor dopant makes the semiconducting ceramics to an insulating one. Barium titanate ceramics have application in various engineering fields.

2.1.1 Multilayer capacitor (MLCs)

Mostly $BaTiO_3$ is used in capacitor due to having a high dielectric constant. In thin dielectric form packed in a minimum space not only with high capacitance but also mechanically tough [27]. It is used as passive component in the circuit for large scale integration (LSI). It is also cost effective one with the use of internal electrode made of, nickel (Ni) and copper (Cu) [28].

2.1.2 (s2) PTC thermistor

The high resistivity of $BaTiO_3$ makes it as good candidates for PTC thermistor. Doping at A and B site of $BaTiO_3$ convert them as a semiconducting material. Below the Curie temperature these semiconductor materials have low resistance. Above crystal particle boarder barrier layer persuades by the surface state. So FE characteristics with high dielectric constant exhibit by this crystal boundary of high resistance. So below Curie temperature the potential barrier is small. Low resistivity has been exhibited in these materials due to effortlessly penetration of electron. The height of potential barrier increases above the Curie temperature makes electron difficult to pass through it results in increasing the resistivity of the material.
In various electronics circuits this semiconducting $BaTiO_3$ materials are used as constant temperature heaters or switching devices. For temperature or parameter related to temperature can be detected, measured and control of temperature with

the use of this PTC thermistor. Among all the available sensor materials PTC has the highest temperature coefficient of resistance.

2.1.3 Nanogenerator (NG)

High piezoelectric effect with biocompatibility make them use as Nanogenerator. Piezoelectric potential is induced in between the two electrodes of NG, this is due to mechanical stress. Commercial devices can be work using this generated electrical energy without an external energy source.

2.2 Strontium titanate ($SrTiO_3$)

Another member of perovskite family is strontium ferrite STO ($SrTiO_3$) is a complex oxide. It exhibits cubic structure at room temperature. O^{2-} ions are bonded with six folded coordinate to Ti^{4+} ions and with twelve folded coordinate to Sr^{2+} ion. Each Sr^{2+} ion lies in between four TiO_6 octahedral. To decrease in temperature, it undergoes a phase transition. The first transition below room occur at −168°C. The opposite rotation of adjacent oxygen octahedral at this term turns the cubic structure to tetragonal structure. To further decrease in temperature at −236°C the changeable phonon modes turns the tetragonal structure to orthorhombic structure. The orthorhombic structure is change to rhombohedra structure at −263°C [29]. As Curie–Weiss law suggests the dielectric constant increases with the phase transition below room temperature. At this temperature due to quantum fluctuations leads to the quantum PE [30]. The charge storage capacity is high, chemical stability, optical transparency in the visible region with good insulating properties makes it use in modern electronics applications such as phase shifters, high-voltage capacitors, delay lines, filters, tunable oscillators etc. [31–33]. It is used in cancer treatment and in thermo – electric generators due to they have insolubility and high melting point properties [34, 35]. It shows photoconductivity when exposed to light due to having an direct gap and indirect band gap of 3.75 eV and 3.25 eV respectively [36]. Its conductivity increases with the contact of light to the crystal. Its conductivity persists for several days, with small decay [37, 38].

The first oxide, which is superconducting below 0.35 K is the strontium oxide (STO) [39]. It can be used as a tremendous substrate for superconductors with a high operating temperature and for oxide-based thin films. It is used as single-crystal substrate due to the enhanced electrical conductivity of niobium doped STO for the growth of perovskite oxides. It is also used for other ferroelectric and magnetic oxide substrate due to its large lattice parameter. Variation of temperature and pressure parameter can lead to increase oxygen vacancies in both crystal and thin films of $SrTiO_3$. So it becomes more conducting due to stimulate free electrons in the conduction band and also opaque. This used as gate-dielectric material due to the growth of high quality growth of epitaxial $SrTiO_3$ layers on silicon without forming silicon oxide Furthermore, it allow the incorporation of other perovskite oxides based thin-film on top of silicon [40]. It is also used for its piezoelectricity, ferroelectricity and Pyro electricity application in nanoscale [41–44]. It is used for various technological applications such as in super capacitor, nonvolatile memory, tunable microwave capacitors, ultralow-temperature scanning microscopies, high-density dynamic random access memories, soft phonon devices, oxygen sensors [45, 46].

2.3 Barium strontium titanate (BST)

The model of $BaTiO_3$ (BTO) has been used to develop Barium strontium titanate ($Ba_xSr_{1-x}TiO_3$ perovskite. This perovskite also undergoes phase change at it

Curie temperature depend upon the ratio of Ba:Sr. If the ratio decreases the curie temperature decreases. The dielectric constant at the Curie temperature of BST is more than BTO. BST is simply ferroelectric with spontaneous polarization below the Curie temperature. Nearly T_c the tenability of BST has extremely high in the FE phase. So in this phase find application for non volatile application. Above the Curie temperature the pyroelectric BST finds its applications in tunable microwave device associate with low dielectric loss and high dielectric constant. It also finds application in phase shifters, tunable filters and tunable antennas due to its composition dependent curie temperature with permittivity depends electric field.

Due to its high value of capacitance make it useful to construct high capacitance capacitor. It has uses in tunable microwave devices tunable capacitor, phase shifters, tunable transformers. BST varactors are a good replacement of the presently used semiconductor varactor and mechanical tuners. It is not only the drawback of large size with small tuning speed of mechanical tunners, but also small power handling capability of semiconductor varactor. BST also used in band pass and low pass tunable filter.

Semiconductor based phase shifter are used in fighter aircraft radar and cellular telephone base stations are associated with high loss at microwave frequencies with low power use ability. BST is the best replacement of these semiconductor based phase shifter associate with small loss, inexpensive and with better power handling properties. It is used in micro strip antenna. In tunable microwave application thin film of BST is used. The reduced size with small weight makes it compatible with microwave circuit. Dielectric constant not decreases sharply with the variation of the thickness of the film [47–49]. When it's used in metal–insulator–metal capacitor shows high dielectric constant, low dielectric loss, high leakage current density, high charge storage density makes it's used in dynamic random access memory (DRAM).

2.4 Lead titanate ($PbTiO_3$)

Another member of perovskite family is the inorganic lead salt of titanic acid compound, i.e. Lead titanate ($PbTiO_3$). Yellow powder of lead titanate is water insoluble. It shows a high Curie point of 490°C. It shows second order phase transition due to which it changes from cubic pyroelectric phase to ferroelectric tetragonal phase. At room temperature it shows the tetragonal structure belongs to *P*4 mm space group. It undergoes large volume change when cooled below the Curie temperature. It is not easy to formulate it in the bulk form. There is formation of crack during manufacture due to strain. To reduce this strain various dopants are used to modify the lead titanate. Ferroelctric lead titanate find its uses in, resonators, actuators, IR sensors, ultrasonic transducers and MLCs etc. [50–52]. Various process such as melting, Co precipitation, decomposition, hydrothermal, sol–gel, chemical vapor deposition, molecular beam epitaxy (MBE), molten salt methods, solid state method and sputtering is used to prepare thin films, single crystals and ceramic powders of $PbTiO_3$ [53–74]. For advance electromechanical devices, it is the capable building blocks. It can be prepared in micro tube, Nano sized powder and nanowire by the use of hydrothermal method [68, 75–79]. It is one of the most used materials for the fabrication of sensors, memory capacitors, optoelectronics devices etc. [76, 79].

2.5 Lead zirconate titanate (PZT)

Lead zirconate titanate with perovskite structures is an inorganic intermetallic compound. It is a solid solution of lead zirconate and lead titanate represented as [Pb (Zr_xTi_{1-x}) O_3, $0 \leq x \leq 1$]. PZT has Pb^{2+} ions at A site with random occupation of B site by Ti^{4+} and Zr^{4+} ions. It undergoes phase change with composition, but also with temperature. Above its Curie temperature it shows the pyroelectric effect with cubic structure. It undergoes a structural change from PE cubic phase either to FE rhombohedral phase or tetragonal phase. The spontaneous polarization in these structures is oriented in< 100⟩ and ⟨111⟩ set of directions for tetragonal and rhombohedral phase respectively. The morphotropic phase boundary is around at 52/48 of Zr/Ti ratio separating FE tetragonal and orthorhombic phases. At this boundary PZT has shown maximum dielectric and piezoelectric constants. Rhombohedral phase with 8 possible domains and tetragonal phase with 6 domains with total 14 domains are equally favorable at this composition. Piezoelectric PZT ceramics can be tailored according to application with ions having valence diverse from the host ions in the lattice. Especially PZT at MPB is modified to form soft and hard PZT. Acceptor ions are used either by the use of Al^{3+}, Fe^{3+} at B site and Na^{+} and K^{+} at A site of the perovskite to produce oxygen vacancies in the lattice [80, 81]. Donor ions PZT produced domains wall motion fromed by Nd^{3+}, La^{3+} at A site where as Nb^{5+}, Sb^{5+} at B site in the lattice [82–85]. Hard PZT formed by donor doping shows low dielectric constant, small electrical losses, small piezoelectric co-efficient with high coercive field. So it is difficult to pole and dipole the sample and make them useful for rough applications. Whereas soft PZT has shown larger losses, high dielectric constant and piezoelectric co-efficient. So they can easily dope and dipole. They are used in FeRAM, actuator of STM/AFM type, ultrasound transducer, capacitor with high dielectric constant, IR sensor, etc. High value of piezoelectric coefficient of FE PZT make them useful for micro sensor, micro actuator used micro electromechanical system (MEMS) devices [86–88]. Nano rods, wire and hollow tube of PZT are now used for different application [89–98]. PZT Nanogenerator are used for piezoelectric effect to produce piezoelectric energy in the microscale. The intrinsic polar crystal structure of PZT nanofibers shows high piezoelectric voltage constant. High aspect ratio of nanostructure PZT overcomes extremely brittle nature observed in bulk PZT and its thin films and used for generation of energy in alternating loads [89, 90].

2.6 Lead magnesium niobate (PMN)

Single crystal of Pb ($Mg_{1/3}Nb_{2/3}$) O_3 (PMN) materials belongs to the FE relaxor materials and are used in [99], and the work in the electrostrictive actuators with high-strain and capacitor with high dielectric constant [100–102]. The Perovskite type structure of Pb (B_1, B_2) O_3 formula with lower valence B_1 such as Zn^{2+}, Mg^{2+}, Fe^{3+} and Ni^{2+} and higher valency B_2 is such as Ta^{5+}, Nb^{5+} and W^{5+}. On cooling below the Curie temperature relaxor PMN ferroelectrics shows a wide dispersive and the diffused phase transition. Heterogeneity in composition on a microscopic scale leads to the diffuse phase transition in the relaxor ferroelectric.

In micoregion the stoichiometry is not obeyed by disorder B-site leads to change in FE transition temperatures leads to broadening dielectric peak. Strong frequency dependent dielectric constant has been observed in the relaxor ferroelectric. Curie temperature changes linearly with frequency with high dielectric loss below this temperature. Some relaxor shows second order phase transition. This relaxor remnant polarization is starting to decrease from Curie temperature and becomes zero with the increasing in temperature [103–106].

2.7 Potassium niobate ($KNbO_3$)

KN ($KNbO_3$) is also exhibited perovskite structure. It shows interesting ferroelectric properties at low temperature. It shows three phase transitions simple displacive type at low temperature with different symmetry and PE phase at high-temperature [107]. Oxygen behaves as rigid body in the octahedral and vibrates about Nb atoms [108]. Liberation of oxygen octahedral leads to the irregular anisotropy exhibited by oxygen atom due to mean square displacements. Its structure shows two subshell obtained from the splitting of oxygen octahedron. Six niobium atoms are from the third nearest sub shell. 24 oxygen atoms form a fourth adjacent octahedral shell consists of four sub shell of six atoms. Fifth shell is made of 12 niobium atoms. Neutron diffraction study predicts the structural change with temperature $KNbO_3$. It shows three phase transition from cubic to tetragonal, tetragonal–orthorhombic and orthorhombic–rhombohedral at $T \approx 418°C$, 225°C and 10°C respectively [109, 110], Transverse optic mode exhibited by $KNbO_3$ is softened with lessening temperature obtain from Raman, IR and inelastic neutron scattering [111–113]. Soften mode frequency obtain from dielectric measurement is good agreement with values calculated Cochran from $\omega^2 \propto (T-T_0$ with $T_0 \approx 370°C$ Curie–Weiss temperature [114]. $KNbO_3$ Curie–Weiss constant was found to be about 2.8×10^5 K. Displacive model is used to calculate this shows good agreement with the theoretical value. Large electromechanical coupling factor with zero temperature coefficients at room temperature exhibited by KN crystal is used for piezoelectric application [115, 116]. Surface acoustic wave (SAW) filter prepared using $KNbO_3$ find its application in mobile phones and television receivers [115, 116]. The crystal symmetry of KN crystal shows 49.5° rotation about the y axis by the x-cut [117, 118]. In high quality fiber shape these crystals show small lattice defects [119]. Different melting temperature hinders to grow high quality and large size KN crystal [115]. Both Bridgman (BM) technique and Top-seeded solution growth (TSSG) method is used, however the Bridgman (BM) technique is the easier one to grow bulk shape KN crystals [120–124]. Phase diagram suggests line compounds are formed when these crystals are developed from high-temperature solutions [125]. A peritectic transformation is shown when it grows from molten stoichiometric composition. KN in nanorod form are used capacitor and nano (NG) [126].

2.8 Sodium niobate ($NaNbO_3$)

This is also the member of perovskite family, but with anti ferroelctric properties. Six phase transition in between −200 to 650°C range affects its structural, dielectric, and optical properties. Phase transition at 200,360 and 480°C has been observed due to off center displacement of Nb ion with tilting of oxygen octahedral. Its stable cubic structure of the *Pm3m* space group has been observed at high temperature, i.e. >640°C [127, 128]. The orthorhombic structure of the phase *Pbcm* space group has been observed at room temperature is antiferroelctric one. At 360°C the antiferroelcetric orthombic structure with *Pbma* space group undergoes phase transition to antiferroelectric orthorhombic structure with a *Pnmm* space group associated with maximum dielectric constant [129–134]. $NaNbO_3$ single crystal exhibit low-frequency relaxation processes [135]. Distinct discontinuity is observed in mean relaxation time and relaxation parameter at Curie temperature Tc. At high temperature, low frequency relaxation increases due to crystalline structure disorder. This leads formation of local dipole in the polar region [136]. Stimulating electrical and mechanical properties of Sodium niobate-based ceramics make them a useful candidate for many technological applications [136–147]. At attainable electric fields this well-known antiferroelectric shows FE properties. Cost effective lead-free nanowire based $NaNbO_3$ piezoelectric has a high-output [148]. So it find application in hologram and optical data storage having high density [137–143]. It also used as nanocapacitors, NGs and in the memories of nanoscale [148–150]. For large-scale lead-free piezoelectric NG may be $NaNbO_3$ nanowires is one of useful candidate [151].

3. Conclusions

Several reports on ferroelectric materials and their use for different piezoelectric application has been studied in the last few years. As discussed in this chapter the effect of perovskite structure affect its ferroelectric properties. Doping in these perovskite structures also responsible for the enhanced of its properties by tailoring its crystal structure. These materials now investigated in the composite, nanowire and the nanorods form to make the device mechanically robust and more compact. This is a vital field for the research, as a key element in the digital world.

Author details

Paramjit Kour[1*] and Sudipta Kishore Pradhan[2]

1 Department of Physics, Birla Institute of Technology, Patna, Bihar, India

2 Department of Mechanical Engineering, Birla Institute of Technology, Patna, Bihar, India

*Address all correspondence to: paramjit.kour@bitmesra.ac.in

References

[1] von Hippel, A., Ferroelectricity, domain structure, and phase transitions of barium titanate, Rev. Mod. Phys., Vol. 22, 222-237 (1950).

[2] Valasek, J. Piezo-electric and allied phenomena in Rochelle salt, Phys. Rev., 17, 475-481 (1921).

[3] W. Kanzig, Ferroelectrics and Antiferroelectrics, Academic Press, New York (1957).

[4] Hao, YM., Lou, SY., Zhou, SM. et al. Structural, optical, and magnetic studies of manganese-doped zinc oxide hierarchical microspheres by self-assembly of nanoparticles. Nanoscale Res Lett 7, 100 (2012).

[5] Kanzig, W. History of ferroelectricity, 1938-1955. Ferroelectrics, 74, 285-291 (1987).

[6] Cross, L.E. and Newnham, R.E. History of Ferroelectrics, Ceramics and Civilization, High-Technology Ceramics—Past, Present, and Future, vol. 3, American Ceramic Society, Westerville, OH, pp. 289-305 (1987).

[7] Busch, G, How i discovered the ferroelectric properties of KH_2PO_4. Ferroelectrics, 71, 43-47 (1987).

[8] Fousek, J., Ferroelectricity: remarks on historical aspects and present trends. Ferroelectrics, 113, 3-20 (1991).

[9] Feldman, C., Formation of thin films of $BaTiO_3$ by evaporation. Review of Scientific Instruments, 26(5), 463-466 (1955).

[10] Haertling, G.H, Ferroelectric ceramics: history and technology. J. Am. Ceram. Soc., 82 (4), 797-818 (1999).

[11] Jin, Wenchao and Wang, Zhao and Huang, Hao and Hu, Xiaokang and He, Yahua and Li, High-performance piezoelectric energy harvesting of vertically aligned $Pb(Zr,Ti)O_3$ nanorod arrays, RSC Adv, 8, 7422-7427 (2018).

[12] Seung, W., Yoon, H.J., Kim, T.Y., Ryu, H., Kim, J., Lee, J.H., Lee, J.H., Kim, S., Park, Y.K., Park, Y.J. and Kim, S.W., Boosting power-generating performance of triboelectric nanogenerators via artificial control of ferroelectric polarization and dielectric properties. Advanced Energy Materials, 7(2), 1600988 (2017).

[13] Hench, L.L. and West, L.K. (1990) Principles of Electronic Ceramics, John Wiley and Sons, Inc., 244-247.

[14] Kingery, W.D., Bowen, H.K., and Uhlmann, D.R. (1976) Introduction to Ceramics, John Wiley and Sons, Inc., pp. 926-927.

[15] Guo, L., Luo, H., Gao, J., Guo, L., and Yang, J. Microwave hydrothermal synthesis of barium titanate powders. Mater. Lett., 60, 3011 (2006).

[16] Arya, P.R., Jha, P., Subbanna, G.N., and Ganguli, A.K. Polymeric citrate precursor route to the synthesis of nano-sized barium lead titanates. Mater. Res. Bull., 38, 617 (2003).

[17] Boulos, M., Guillement-Fritsch, S., Mathieu, F., Durand, B., Lebey, T., and Bley, V. Hydrothermal synthesis of nanosized $BaTiO_3$ powders and dielectric properties of corresponding ceramics. Solid State Ionics, 176, 1301 (2005).

[18] Xu, H. and Gao, L. Tetragonal nanocrystalline barium titanate powder: preparation, characterization, and dielectric properties. J. Am. Ceram. Soc., 86, 203 (2003).

[19] Vinothini, V., Singhand, P., and Balasubramanian, M. Synthesis of barium titanate nanopowder using

polymeric precursor method. Ceram. Int., 32, 99 (2006).

[20] Duran, P., Gutierrez, D., Tartaj, J., and Moure, C. Densification behaviour, microstructure development and dielectric properties of pure $BaTiO_3$ prepared by thermal decomposition of (Ba, Ti)-citrate polyester resins. Ceram. Int., 28, 283 (2002).

[21] Stojanovic, B.D., Jovalekic, C., Vukotic, V., Simoes, A.Z., and Varela, J.A. Ferroelectric properties of mechanically synthesized nanosized barium titanate. Ferroelectrics, 319, 65 (2005).

[22] Buscaglia, V., Viviani, M., Buscaglia, M.T., Nanni, P., Mitoseriu, L., Testino, A., Stytsenko, E., Daglish, M., Zhao, Z., and Nygren, M. Nanostructured barium titanate ceramics. Powder Technol., 148, 24 (2004).

[23] Simon-Seveyrat, L., Hajjaji, A., Emziane, Y., Guiffard, B., and Guyomar, D. Re-investigation of synthesis of $BaTiO_3$ by conventional solid-state reaction and oxalate coprecipitation route for piezoelectric applications. Ceram. Int., 33, 35 (2007).

[24] Kim, H.T. and Han, Y.H. Sintering of nanocrystalline $BaTiO_3$. Ceram. Int., 30, 1719 (2004).

[25] Benlahrache, M.T., Barama, S.E., Benhamla, N., and Achour, A. Influence of polarization electric field on the dielectric properties of $BaTiO_3$-based ceramics. Mater. Sci. Semicond. Process., 9, 1115 (2006)

[26] Buscaglia, M.T., Buscaglia, V., Viviani, M., Nanni, P., and Hanuskova, M. Influence of foreign ions on the crystal structure of $BaTiO_3$. J. Eur. Ceram. Soc., 20, 1997 (2000).

[27] Wang, H.-L. Structure and dielectric properties of perovskite-barium titanate ($BaTiO_3$). MatE, 115 (2002).

[28] Kuo, D.-H., Wang, C.-H., and Tsai, W.-P. Donor- and acceptor-cosubstituted $BaTiO_3$ for nonreducible multilayer ceramic capacitors. Ceram. Int., 32, 1-5 (2006).

[29] Lytle, F.W. X-ray diffractometry of low-temperature phase transformations in strontium titanate. J. Appl. Phys., 35, 2212 (1964).

[30] Muller, K.A. and Burkard, H. $SrTiO_3$: an intrinsic quantum paraelectric below 4 K. Phys. Rev. B, 19 (7), 3593-3602 (1979).

[31] Kolodiazhnyi, T. and Petric, A. The applicability of Sr-deficient n-type $SrTiO_3$ for SOFC anodes. J. Electroceram., 15 (1), 5-11 (2005).

[32] Rehn, L.E. Ion beams in high-temperature superconductivity research. Nucl. Instrum. Methods Phys. Res., Sect. B, 64, 161 (1992).

[33] Ramirez, P. Colossal magnetoresistance. J. Phys.: Condens. Matter, 9, 8171-8199 (1997).

[34] Power Sources for Remote Arctic Applications, U.S. Congress, Office of Technology Assessment (1994), Washington, DC, OTA-BP-ETI-129.

[35] Standring, W.J.F., Selnæs, Q.G., Sneve, M., Finne, I.E., Hosseini, A., Amundsen, I., and Strand, P. (2005) Assessment of Environmental, Health and Safety Consequences of Decommissioning Radioisotope Thermal Generators (RTGs) in Northwest Russia.

[36] Van Benthem, K., Elsässer, C., and French, R.H. Bulk electronic structure of $SrTiO_3$: experiment and theory. J. Appl. Phys., 90, 6156 (2001).

[37] Marianne C. Tarun, Farida A. Selim, and Matthew D. McCluskey Persistent Photoconductivity in Strontium Titanate, Phys. Rev. Lett., 111, 187403 (2013).

[38] Tamarra Kemsley, Light Exposure Increases Crystal's Electrical Conductivity 400-fold, Nature World News, Retrieved 18 November (2013).

[39] Koonce, C.S. and Cohen, M.L. Superconducting transition temperatures of semiconducting $SrTiO_3$. Phys. Rev., 163 (2), 380 (1967).

[40] McKee, R.A., Walker, F.J., and Chisholm, M.F., Crystalline oxides on silicon: the first five monolayers. Phys. Rev. Lett., 81 (14), 3014 (1998).

[41] Urban, J.J., Yun, W.S., Gu, Q., and Park, H. Synthesis of single crystalline nanorods composed of Barium titanate and Strontium titanate. J. Am. Chem. Soc., 124 (7), 1186-1187 (2002).

[42] Márquez-Herrera, A., Ovando-Medina, V.M., Corona-Rivera, M.A., Hernandez-Rodriguez, E., Zapata-Torres, M., Campos-Gonzalez, E., Guillen-Cervantes, A., Zelaya-Angel, O., and Meléndez-Lira, M. A novel solvothermal route for obtaining strontium titanate nanoparticles. J. Nanopart. Res., 15, 1525 (2013).

[43] Kiat, J.M., Bogicevic, C., Gemeiner, P., Al-Zein, A., Karolak, F., Guiblin, N., Porcher, F., Hehlen, B., Yedra, L.L., Estradé, S., Peiró, F., and Haumont, R., Structural investigation of strontium titanate nanoparticles and the core-shell model. Phys. Rev. B, 87, 024106 (2013).

[44] Goto, T., Nakashima, K., Fujii, I., Kuroiwa, Y., Makita, Y., Ryu, M., Suzuki, T., and Wada, S. Preparation of barium titanate/strontium titanate multilayer complex nanoparticles using nanocube substrates. IOP Conf. Ser. Mater. Sci. Eng., 18, 092030 (2011).

[45] Voigts, F., Damjanovic, T., Borchardt, G., Argirusis, C., and Maus-Friedrichs, WSynthesis and characterization of strontium titanate nanoparticles as potential high temperature oxygen sensor material. J. Nanomater., 6, 63154 (2006).

[46] AZO Nano Strontium Titanate ($SrTiO_3$) Nanoparticles-Properties and Applications (2016).

[47] Shaw, T.M., Suo, Z., Huang, M., Liniger, E., Laibowitz, R.B., and Baniecki, J.D. The effect of stress on the dielectric properties of barium strontium titanate thin films. Appl. Phys. Lett., 75, 2129-2131 (1999).

[48] Maria, J.-P., Parker, C.B., Kingon, A.I., and Stauf, G. Thickness, strain, and temperature-dependent properties of barium strontium titanate thin films. IEEE International Symposium on Applications of Ferroelectrics, pp. 151-154 (2002).

[49] Basceri, C., Streiffer, S.K., Kingon, A.I., and Waser, R. The dielectric response as a function of temperature and film thickness of fiber-textured (Ba, Sr)TiO_3 thin films grown by chemical vapor deposition. J. Appl. Phys., 82 (5), 2497-2504 (1997).

[50] Suchicital, C.T.A. and Payne, D.A. Flux growth of single crystal lead titanate. J. Cryst. Growth, 104, 211 (1990).

[51] Jaffe, B., Cook, W.R. Jr., and Jaffe, H. (1971) Piezoelectric Ceramics, Academic Press, London, New York.

[52] Meyer, R. Jr., Newnham, R., Alkoy, S., Ritter, T., and Cochran, J. Jr., Pre-focused lead titanate >25 MHz single-element transducers from hollow spheres. IEEE Trans. Ultrason. Ferroelectr. Freq. Control, 48 (2), 488-493 (2001).

[53] Kobayashi, J. Growing of ferroelectric $PbTiO_3$ crystals. J. Appl. Phys., 29, 866 (1958).

[54] Sun, B.N., Huang, Y., and Payne, D.A. Growth of large $PbTiO_3$ crystals by a self-flux technique. J. Cryst. Growth, 128, 867 (1993).

[55] Ganesa Moorthy, S., Joseph Kumar, F., Balakumar, S., and Subramanian, C. Growth of lead titanate single crystals and their characterization. J. Korean Phys. Soc., 32, S1214–S1216 (1998).

[56] Grabmaier, B.C. $PbTiO_3$ grown from the melt. Ferroelectrics, 13, 501 (1976).

[57] Oka, K., Unoki, H., Yamaguchi, H., and Takahashi, H. Crystal growth of $PbTiO_3$ by the top-seeded solution-growth method. J. Cryst. Growth, 166, 380 (1996).

[58] Blum, J.B. and Gurkovich, S.R. Sol-gel-derived $PbTiO_3$. J. Mater. Sci., 20, 4479 (1985).

[59] Tangboriboon, N., Jamieson, A.M., Sirivat, A., and Wongkasemjit, S. A novel route to perovskite lead titanate from lead and titanium glycolates via the sol–gel process. Appl. Organomet. Chem., 20 (12), 886-894 (2006).

[60] Lanki, M., Nourmohammadi, A., and Feiz, M.H. (2012) Lead partitioning in sol–gel derived lead titanate nanopowders. Applications of Ferroelectrics Held Jointly with 2012 European Conference on the Applications of Polar Dielectrics and 2012 International Symposium Piezoresponse Force Microscopy and Nanoscale Phenomena in Polar Materials (ISAF/ECAPD/PFM), 2012 International Symposium, 2012, pp. 1-4.

[61] Fox, G.R., Adair, J.H., and Newnham, R.E. Effects of pH and H_2O_2 upon co precipitated $PbTiO_3$ powders. J. Mater. Sci., 25, 3634 (1990).

[62] Fang, J., Wang, J., Ng, S.C., Chew, C.H., and Gan, L.M., Preparation and characterisation of ultrafine lead titanate ($PbTiO_3$) powders. J. Mater. Sci., 34, 1943-1952 (1999).

[63] Fang, J., Wang, J., Gan, L.-M., and Ng, S.-C., Comparative study on phase development of lead titanate powders. Mater. Lett., 52, 304-312 (2002).

[64] Kim, M.J. and Matijevic, E. Preparation and characterization of ′ uniform submicrometer lead titanate particles. Chem. Mater., 1, 363 (1989).

[65] Watson, D.J., Randall, C.A., Newnham, R.E., and Adair, J.H. (1988) in Ceramics Transactions, Ceramic Powder Science II, vol. 1 (eds G.L. Messing, E.R. Fuller Jr., and H. Hausner), The American Ceramic Society Inc., Westerville, OH, p. 154.

[66] Blakea, G.R., Armstronga, A.R., Sastreb, E., Zhoua, W., and Wrighta, P.A. The preparation of a novel layered lead titanate and its conversion to the perovskite lead titanate $PbTiO_3$. Mater. Res. Bull., 36, 1837-1845 (2001).

[67] Morita, T. and Cho, Y. A hydrothermally deposited epitaxial lead titanate thin film on strontium ruthenium oxide bottom electrode. Appl. Phys. Lett., 85, 2331 (2004).

[68] Morita, T. and Cho, Y. Piezoelectric property of an epitaxial lead titanate thin film deposited by the hydrothermal method. Appl. Phys. Lett., 88, 112908 (2006).

[69] Yoon, S.-.G., Park, J.-.D., Choi, J.-.H., and Kim, H.-.G. Preparation, properties, and characterization of thin ferroelectric films of lead titanate. J. Vac. Sci. Technol., A, 9, 281 (1991).

[70] Yoon, S.G. and Kim, H.G. Characterization and electrical properties of chemical vapour deposited ferroelectric lead titanate films on titanium. IEEE Trans. Ultra son. Ferroelectric. Freq. Control, 37 (5), 333-338 (1990).

[71] Rispens, G. and Noheda, B. (2007) Ultra-thin lead titanate films grown by molecular beam epitaxy. Integr. Ferroelectric., 92, 30-39 (2006).

[72] Jacobsen, H., Jung, T., Ortner, K., Schiffmann, K.I., Quenzer, H.V., and

Wagner, B. Development of a piezoelectric lead titanate thin film process on silicon substrates by high rate gas flow sputtering. Sens. Actuators, A, 133 (1), 250-258 (2007).

[73] Arendt, R.H. and Rosolowski, J.H. (1979) Molten salt synthesis of lead zirconate titanate solid solution powder. US Patent 4 152 282, May 1, 1979.

[74] Shrout, T.R., Papet, P., Kim, S., and Lee, G.-S. Conventionally prepared submicrometer lead-based perovskite powders by reactive calcination. J. Am. Ceram. Soc., 73 (7), 1862 (1990).

[75] Zhao, L., Steinhart, M., Yu, J., and Gösele, U. Lead titanate nano- and microtubes. J. Mater. Res., 21 (3), 685-690 (2006).

[76] Hu, Y., Gu, H., Sun, X., You, J., and Wang, JPhotoluminescence and Raman scattering studies on $PbTiO_3$ nanowires fabricated by hydrothermal method at low temperature. Appl. Phys. Lett., 88 (19), 193120 (2006).

[77] Shimada, T., Tomoda, S., and Kitamura, T. Ab initio study of ferroelectricity in edged $PbTiO_3$ nanowires under axial tension. Phys. Rev. B, 79, 024102 (2009).

[78] Leong, M., Bayerl, D.J., Shi, J., and Wang, X. Evolution of lead titanate nanostructures from nanoparticle self-assembly. Sci. Adv. Mater., 4, 832-836 (2012).

[79] Nwe, H.H., Maung, Y.M., Win, T.T., and Kyaw Soe, K.K. Hydrothermal synthesis of nano-sized $PbTiO_3$ powder and epitaxial film for memory capacitor application. Am. J. Mater. Sci. Technol., 1, 22-27 (2012).

[80] Weston, T.B., Webster, A.H., and McNamara, V.M., Lead zirconate-lead titanate piezoelectric ceramics with iron oxide additions. J. Am. Ceram. Soc., 52, 253 (1969).

[81] Uchida, N. and Ikeda, T. Studies on $Pb(Zr\text{-}Ti)O_3$ ceramics with addition of Cr_2O_3. Jpn. J. Appl. Phys., 6, 1292 (1967).

[82] Kulcsar, F. Electromechanical properties of lead titanate zirconate ceramics modified with certain three-or five-valent additions. J. Am. Ceram. Soc., 42, 343 (1959).

[83] Kulcsar, F. Electromechanical properties of lead titanate zirconate ceramics modified with tungsten and thorium. J. Am. Ceram. Soc., 48, 54 (1965).

[84] Atkin, R.B., Holman, R.L., and Fularth, R.M. Substitution of Bi and Nb ions in lead zirconate-titanate. J. Am. Ceram. Soc., 54, 113 (1971).

[85] Banno, H. and Tsunooka, T. Piezoelectric properties and temperature dependences of resonant frequency of WO_3-MnO_2-modified ceramics of $Pb(Zr\text{-}Ti)O_3$. Jpn. J. Appl. Phys., 6, 954 (1967).

[86] Polla, D.L. and Francis, L.F. Ferroelectric thin films in micro-electromechanical systems applications. MRS Bull., 21, 59-65 (1996).

[87] Zhang, Q.Q., Gross, S.J., Tadigadapa, S., Jackson, T.N., Djuth, F.T., and Trolier-McKinstry, S. Lead zirconate titanate films for d33 mode cantilever actuators. Sens. Actuators, A, 105, 91-97 (2003).

[88] Kim, J.H., Wang, L., Zurn, S.M., Li, L., Yoon, Y.S., and Polla, D.L. Fabrication process of PZT piezoelectric cantilever unimorphs using surface micromachining. Integr. Ferroelectric. 15, 325-332 (1997).

[89] Swallow, L.M., Luo, J.K., Siores, E., Patel, I., and Dodds, D. A piezoelectric fibre composite based energy harvesting device for potential wearable applications. Smart Mater. Struct., 17 (2), 025017 (2008).

[90] Chen, X., Xu, S., Yao, N., and Shi, Y., 1.6 V nanogenerator for mechanical energy harvesting using PZT nanofibres. Nano Lett., 10 (6), 2133-2137 (2010).

[91] Xu, S., Hansen, B.J., and Wang, Z.L., Piezoelectric nanowire-enabled power source for driving wireless microelectronics. Nat. Commun., 93, 1-5 (2010).

[92] Wu, W., Bai, S., Yuan, M., Qin, Y., Wang, Z.L., and Jing, T., Lead zirconate titanate nanowire textile nanogenerator for wearable energy-harvesting and self-powered devices. ACS Nano, 6 (7), 6231-6235 (2012).

[93] Gu, L., Cui, N., Cheng, L., Xu, Q., Bai, S., Yuan, M., Wu, W., Liu, J., Zhao, Y., Ma, F., Qin, Y., and Wang, Z.L. Flexible fiber nanogenerator with 209 V output voltage directly powers a light-emitting diode. Nano Lett., 13 (1), 91-94 (2013).

[94] Zhu, G., Wang, A.C., Liu, Y., Zhou, Y., and Wang, Z.L. Functional electrical stimulation by nanogenerator with 58 V output voltage. Nano Lett., 12, 3086-3090 (2012).

[95] Lee, K.Y., Kumar, B., Seo, J.S., Kim, K.H., Sohn, J.I., Cha, S.N., Choi, D., Wang, Z.L., and Kim, S.W. P-type polymer-hybridized high-performance piezoelectric nanogenerators. Nano Lett., 12, 1959-1964 (2012).

[96] Park, K.I., Jeong, C.K., Ryu, J., Hwang, G.T., and Lee, K.J. Flexible and large-area nanocomposite generator based on lead zirconate titanate particles and carbon nanotubes. Adv. Energy Mater., 3, 1539-1544 (2013).

[97] Scott, J.F., Fan, H.J., Kawasaki, S. et al. Terahertz emission from tubular $Pb(Zr, Ti)O_3$ nanostructures. Nano Lett., 8 (12), 4404-4409 (2008).

[98] Bai, S., Xu, Q., Gu, L., Ma, F., Qin, Y., and Wang, Z.L. (2012) Single crystalline lead zirconate titanate (PZT) nano/micro-wire based self-powered UV sensor. Nano Energy, 1 (6), 789-795.

[99] Smolenskii, G.A. and Agranovskaya, A.I. Dielectric polarization of a number of complex compounds. Sov. Phys. Solid State (Eng. Transl.), 1 (10), 1429-1437 (1960).

[100] Cross, L.E., Jang, S.J., and Newnham, R.E. Large electrostrictive effects in relaxor ferroelectrics. Ferroelectrics, 23, 187-192 (1980).

[101] Nomura, S. and Uchino, K. Recent applications of PMN-based electrictors. Ferroelectrics, 50, 197-202 (1983).

[102] Kahn, M., Burks, D., Burn, I., and Schulze, W. in Electronic Ceramics (ed. L.M. Levinson), Marcel Dekker, New York, pp. 191-274 (1988).

[103] Cross, L.E. Relaxor ferroelectrics. Ferroelectrics, 76, 241 (1987).

[104] Smolenski, G.A. and Agranovskaya, A.I. Dielectric polarization and losses of some complex compounds. Sov. Phys. Tech. Phys., 3, 1380 (1958).

[105] Shrout, T.R. and Halliyal, A. Preparation of lead-based ferroelectric relaxors for capacitors. Am. Ceram. Soc. Bull., 66, 704 (1987).

[106] Shrout, T.R. and Dougherty, J.P. Lead Based $Pb(B_1B_2)O_3$ Relaxors vs $BaTiO_3$ Dielectrics for Multilayer Capacitors, Ceramic Transactions, Ceramic Dielectrics: Composition, Processing, and Properties, J. Am. Ceram. Soc., 8, 3 (1990).

[107] Matthias, B.T., New ferroelectric crystals. Phys. Rev. B, 75, 1771 (1949).

[108] Hewat, A.W. Cubic-tetragonal-orthorhombic-rhombohedral ferroelectric transitions in perovskite potassium niobate: neutron powder

profile refinement of the structures. J. Phys. C: Solid State Phys., 6, 2559-2572 (1973).

[109] Frenkel, A.I., Stern, E.A., and Yacoby, Y., Pressure-induced changes in the local structure of $KNbO_3$. AIP Conference Proceedings, 436, 238 (1998).

[110] Perry, C.H., Hayes, R.R., and Tornberg, N.E. (1971) in Proceedings of the 2nd International Conference on Light Scattering of Solids (ed. M. Balkanski), Paris: Flammarion Sciences.

[111] Fontana, M.D., Metrat, G., Servoin, J.L., and Gervais, F. Soft ferroelectric mode in $KNbO_3$. Ferroelectrics, 38, 797 (1981).

[112] Fontana, M.D., Kugel, G.E., Vamvakas, J., and Carabatos, C. Persistence of tetragonal raman lines in cubic $KNbO_3$. Solid State Commun., 45, 873 (1983).

[113] Fontana, M.D., Dolling, G., Kugel, G.E., and Carabatos, C. Inelastic neutron scattering in tetragonal $KNbO_3$. Phys. Rev. B, 20, 3850 (1979).

[114] Jona, F. and Shirane, G. (1962) Ferroelectric Crystals, MacMillan, New York.

[115] Yamanouchi, K., Wagatsuma, Y., Odagawa, H., and Cho, Y. Single crystal growth of $KNbO_3$ and application to surface acoustic wave devices. J. Eur. Ceram. Soc., 21, 2791-2795 (2001).

[116] Yamanouchi, K. and Odagawa, H. Research of super-high electromechanical coupling surface acoustic wave substrates. Jpn. J. Appl. Phys., 40, 3726-3728 (2001).

[117] Nakamura, K. and Kawamura, Y. Orientation dependence of electromechanical coupling factors in $KNbO_3$. IEEE Trans. Ultrason. Ferroelectr. Freq. Control, 47, 750-755 (2000).

[118] Kimura, H., Tanahashi, R., Maiwa, K., Baba, H., Cheng, Z.X., and Wang, X.L. Potassium-sodium-rubidium niobate single crystals and electric properties. Int. J. Mod. Phys. B, 23, 3631-3636 (2009).

[119] Takagi, T., Fujii, T., and Sakabe, Y. Growth and characterization of $KNbO_3$ by vertical Bridgman method. J. Cryst. Growth, 259, 296-301 (2003).

[120] Fukuda, T. and Uematsu, Y. Preparation of $KNbO_3$ single crystal for optical applications. Jpn. J. Appl. Phys., 11, 163-169 (1972).

[121] Fukuda, T., Uematsu, T., and Ito, T. Kyropoulos growth and perfection of $KNbO_3$ single crystal. J. Cryst. Growth, 24-25, 450-453 (1974).

[122] Xing, W., Looser, H., Wuest, H., and Arend, HProgress in $KNbO_3$ crystal growth. J. Cryst. Growth, 78, 431-437 (1986).

[123] Zgonik, M., Schlesser, R., Biaggio, I., Voit, E., Tscherry, J., and Günter, P. Materials constants of $KNbO_3$ relevant for electro and acousto-optics. J. Appl. Phys., 74, 1287-1297 (1993).

[124] Wiesendanger, E. Dielectric, mechanical and optical properties of orthorhombic $KNbO_3$. Ferroelectrics, 6, 263-281 (1974).

[125] Reisman, R. and Holtzberg, F. Phase equilibria in the system K_2CO_3-Nb_2O_5 by the method of differential thermal analysis. J. Am. Chem. Soc., 77, 2115-2119 (1955).

[126] Jung, J.H., Chen, C.-Y., Yun, B.K., Lee, N., Zhou, Y., Jo, W., Chou, L.-J., and Wang, Z.L. Lead-free $KNbO_3$ ferroelectric nanorod based flexible nanogenerators and capacitors. Nanotechnology, 23, 375401 (2012).

[127] Mishra, S.K., Choudhury, N., Chaplot, S.L., Krishna, P.S.R., and

Mittal, R. Competing antiferroelectric and ferroelectric interactions in $NaNbO_3$: neutron diffraction and theoretical studies. Phys. Rev. B, 76, 024110 (2007).

[128] Mishra, S.K., Mittal, R., Pomjakushin, V.Y., and Chaplot, S.L. Phase stability and structural temperature dependence in sodium niobate: a high-resolution powder neutron diffraction study. Phys. Rev. B, 83, 134105 (2011).

[129] Barth, T. (1925) Die Kristallstruktur von Perowskit und verwandten Verbindungen. Nor. Geol. Tidsskr., 8, 201.

[130] Cross, L.E. and Nicholson, B.J. The optical and electrical properties of single crystals of sodium niobate. Philos. Mag., 46, 453-466 (1955).

[131] Ahtee, M., Glazer, A.M., and Megaw, H.D. The structures of sodium niobate between 480 and 575°C and their relevance of soft-phonon modes. Philos. Mag., 26 (4), 995-1014 (1972).

[132] Lefkowitz, I., Lukaszewicz, K., and Megaw, H.D. The high temperature phase of sodium niobate and the nature of transitions in pseudosymmetric structure. Acta Crystallogr., 20, 670-683 (1966).

[133] Ishida, K. and Honjo, G. Soft mode and super lattice structure in $NaNbO_3$. J. Phys. Soc. Jpn., 34 (5), 1279-1288 (1973).

[134] Glazer, A.M. and Megaw, H.D. Studies of the parameters and domains in the phase transition of $NaNbO_3$. Acta Crystallogr., A29, 489-495. (1973)

[135] Konieczny, K. and Kajtoch, C. Low-frequency dielectric dispersion in $NaNbO_3$ single crystals. Ferroelectrics, 215, 65-73 (1998).

[136] Konieczny, K. Dielectric relaxation in $NaNbO_3$ single crystal. Condens. Matter Phys., 2 (4), 655-660 (1999).

[137] Xu, Y. (1991) Ferroelectric Materials and Their Applications, North Holland, Amsterdam.

[138] Lines, M.E. and Glass, A.M. (1977) Principles and Applications of Ferroelectrics and Related Materials, Clarendon Press, Oxford.

[139] Tejuca, L.G. and Fierro, J.L.G. (1993) Properties and Applications of Perovskite-Type Oxides, Marcel Dekker, New York. 176

[140] Valdez, E., de Araujo, C.B., and Lipovskii, A.A. Second harmonic scattered light from a transparent glass-ceramic containing sodium niobate nanocrystals. Appl. Phys. Lett., 89, 031901 (2006).

[141] Hollenstein, E., Davis, M., Damjanovic, D., and Setter, N. Piezoelectric properties of Li- and Ta-modified $(K_{0.5}Na_{0.5})$ NbO_3 ceramics. Appl. Phys. Lett., 87, 182905 (2006).

[142] Falcão-Filho, E.L., Bosco, C.A.C., Maciel, G.S., Acioli, L.H., de Araújo, C.B., Lipovskii, A.A., and Tagantsev, D.K. Third-order optical nonlinearity of a transparent glass ceramic containing sodium niobate nanocrystals. Phys. Rev. B, 69, 134204 (2004).

[143] Maciel, G.S., Rakov, N., de Araujo, C.B., Lipovskii, A.A., and Tagantsev, D.K. Optical limiting behavior of a glass–ceramic containing sodium niobate crystallites. Appl. Phys. Lett., 79 (5), 584 (2001).

[144] Saito, Y., Takao, H., Tani, T., Nonoyama, T., Takatori, K., Homma, T., Nagaya, T., and Nakamura, M. Lead-free piezoceramics. Nature, 432, 84 (2004).

[145] Cross, E. Materials science: lead-free at last. Nature, 432, 24 (2004).

[146] Yuzyuk, Y.I., Simon, P., Gagarina, E., Hennet, L., Thiaudiere, D.,

Torgashev, V.I., Raevskya, S.I., Raevskii, I.P., Reznitchenko, L.A., and Sauvajol, J.L. Modulated phases in $NaNbO_3$: Raman scattering, synchrotron x-ray diffraction, and dielectric investigations. J. Phys. Condens. Matter, 17, 4977 (2005).

[147] Yuyuk, Y.I., Gagarina, E., Simon, P., Reznitchenko, L.A., Hennet, L., and Thiaudiere, D. Synchrotron x-ray diffraction and Raman scattering investigations of $(Li_xNa_{1-x})NbO_3$ solid solutions: evidence of the rhombohedral phase. Phys. Rev. B, 69, 144105 (2004).

[148] Xu, S., Yeh, Y.-W., Poirier, G., McAlpine, M.C., Register, R.A., and Yao, N. Flexible piezoelectric PMN-PT nanowire-based nanocomposite and device. Nano Lett., 13, 2393-2398 (2013).

[149] Yan, C., Nikolova, L., Dadvand, A., Harnagea, C., Sarkissian, A., Perepichka, D.F., Xue, D., and Rosei, F. Multiple $NaNbO_3/Nb_2O_5$ heterostructure nanotubes: a new class of ferroelectric/semiconductor nanomaterials. Adv. Mater., 22 (15), 1741-1745 (2010).

[150] Mendoza, M., Khan, M.A.R., Shuvo, M.A.I., Guerrero, A., and Lin, Y., Development of lead-free nanowire composites for energy storage applications., ISRN Nanomater., 151748, 1-8 (2012).

[151] Haertling, G.H. PLZT electrooptic materials and applications—a review. Ferroelectrics, 75, 25 (1987).

Chapter 4

Relaxor Ferroelectric Oxides: Concept to Applications

Lagen Kumar Pradhan and Manoranjan Kar

Abstract

Ferroelectric ceramic is one of the most important functional materials, which has great importance in modern technologies. A ferroelectric ceramic simultaneously exhibits dielectric, piezoelectric, ferroelectric, and pyroelectric properties. The inherent ferroelectric properties are directly related to long-range electric dipoles arrangement in the ferroelectric domains and its response to external stimuli. However, the interruption of the long-range ordering of dipoles leads to the formation of a special class of material is known as relaxor ferroelectric. It shows quite different physical properties as compared to ferroelectric (normal ferroelectric). The origin and design of relaxor ferroelectric are quite interesting for fundamental perspective along with device applications. Therefore, the origin of relaxor ferroelectric along with its fundamental understanding for possible future applications, have been explained briefly in the present chapter.

Keywords: relaxor ferroelectric, spontaneous polarization, ferroelectric domains, polar nanoregions, diffuses phase transition, solid solution

1. Introduction

The origin of ferroelectricity is linked with Rochelle salt and discovered by Joseph Valasek (1897–1993) during the measurement of polarization in Rochelle salt by the applied electric field in 1920 [1]. The fundamental origin of ferroelectricity lies in the response of order parameter (electric dipole) with respect to the applied electric field. However, the field of ferroelectricity remained silent till 1940 [1]. One of the major turning points in ferroelectricity came into the picture around 1940 after the discovery of unusual behavior in dielectric properties of mixed oxides, which are crystallized to perovskite structure [1]. After that, ferroelectricity was intensively focused by the scientific community in all over the world. The perovskite compounds exhibit the ABX_3 structure, where A and B are the cations having different charges and ionic radii. "X" refers to an anion that bonded with both cations. In general, X is often Oxygen (O^{2-}) but also other ions such as sulfides, nitrides, and halides can be considered [2]. However, the perovskite compounds with the general formula ABO_3 create a distinguish place in ferroelectricity compared to others [2]. Currently different groups of perovskite compounds available in the market such as A_2BO_4-layered perovskites (ex: Sr_2RuO_4, K_2NiF_4), $A_2BB'O_6$-double perovskites (ex: Ba_2TiRuO_6) and $A_2A'B_2B'O_9$-triple perovskite (ex: $La_2SrCo_2FeO_9$), etc. [2]. Hence, ferroelectrics are fascinating groups of materials, which have extensively attracted the fundamental understanding of its complex physical properties and emerging device applications (i.e., sensors, actuators, ferroelectric random access memories, etc.) [3]. The presence of spontaneous polarization (P) in ferroelectric material exhibits unique behavior in the presence of external stimuli such as electric field (E), temperature (T), and stress (σ). Nowadays, large numbers of reports are available on the ferroelectric properties of various kinds of materials and their possible applications in modern technologies [3]. Out of the several compounds, Lead Zirconate Titanate ($PbZr_{0.48}Ti_{0.52}O_3$/PZT) is one of the most well-known ferroelectric materials with superior ferroelectric, dielectric, and piezoelectric properties [4]. For a layman's understanding, ferroelectric materials are those which exhibit a high dielectric constant. Furthermore, normal ferroelectrics are characterized by the temperature-dependent maximum dielectric constant near ferroelectric to paraelectric phase transition temperature (T_c) and, also known as Curie's temperature [4]. Some of the basic characteristics of normal ferroelectric are the non-dispersive nature of transition temperature and follow the first-order phase transition. The temperature-dependent dielectric constant of a typical regular ferroelectric ceramic is shown in **Figure 1** [5]. Also, it is well known that the fundamental physics behind the normal ferroelectric lies in the presence of long-range order parameters (electric dipoles) in ferroelectric domains [6].

As per the literature, PZT exhibits the normal ferroelectric in nature with the above-mentioned properties and, well understood experimentally as well as theoretically. However, the spatial substitution of a foreign element such as lanthanum (La^{3+}) in PZT shows the intriguing behavior in terms of dielectric, ferroelectric and piezoelectric properties with respect to frequency, temperature, and electric field, which are different from the normal ferroelectric [7, 8]. The extraordinary properties of modified PZT are related to a special group of ferroelectric materials, which is known as relaxor ferroelectric after the name by the scientist Cross in 1987 [9]. He proposed few characteristic properties for the material to be relaxor ferroelectric, as discussed in the later section. Currently, large numbers of materials with relaxor ferroelectric behavior are available in various forms of crystal structures such as perovskite, layer perovskite, tungsten bronze structure, etc. However, the exact origin of extraordinary properties of relaxor ferroelectrics is still a matter of investigation. A series of explanations have been reported to explain the origin of relaxor behavior by using various models such as

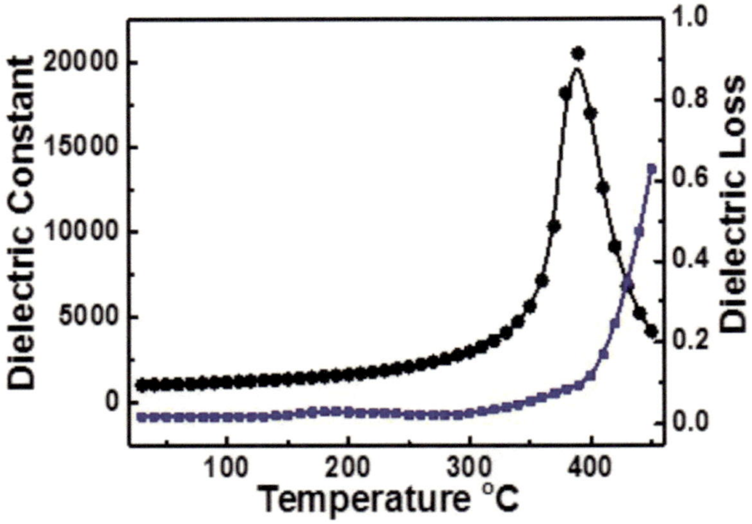

Figure 1.
Temperature-dependent dielectric constant and loss of a typical normal ferroelectric ceramic (PZT ceramic). It is adapted from Ref. [5] (open access).

diffuse phase transition model, dipolar glass model, random field model, super paraelectric model, and so on [10]. In contrast to normal ferroelectrics, the origin of relaxor ferroelectric has been correlated with the presence of compositional fluctuation induced polar nanoregions (PNRs) and tried to explain its properties up to a certain extend [10].

Due to its interesting physical properties, relaxor ferroelectrics can have possible applications in portable electronics, medical devices, pulse power devices, electric vehicles, advanced storage materials, and so on. Therefore, the different aspects from its origin to possible technological applications for future advancement have been discussed briefly in the present chapter.

2. Relaxor ferroelectrics

From the earlier discussion (introduction section), the relaxor ferroelectrics show the abnormal behavior as compared to normal ferroelectric in terms of dielectric, piezoelectric, ferroelectric properties and, subsequently, received much attention by the scientific community. To distinguish a relaxor ferroelectric, Cross assigned few basic properties as follows [9].

1. The temperature-dependent dielectric permittivity (ε') exhibits a broad and smeared maximum, which is known as a diffuse phase transition.

2. The depolarization (T_d) is defined as temperature corresponds to the steepest reduction of remanent polarization.

3. The temperatures (T_m) correspond to maximum dielectric permittivity exhibit strong frequency dependence.

4. There is no macroscopic symmetry breaking (structural changes) near-maximum dielectric temperature (T_m).

5. The Curie–Weiss (C-W) law does not follow the temperature-dependent dielectric permittivity near T_m (dielectric maxima temperature). However, Curie–Weiss law is well fitted above the T_C for normal ferroelectric. Here, T_c is the Curie temperature for normal ferroelectric.

6. Exhibits slim/constricted ferroelectric hysteresis loop due to the presence of nanosize and randomly oriented polar islands known as polar nanoregions (PNRs).

7. Existence of polar nano regions at well above the dielectric maxima temperature 'T_m' up to Burn temperature (T_B), whereas absences in the normal ferroelectrics at above the Curie temperature (T_c).

The temperature-dependent dielectric permittivity of a typical relaxor ferroelectric has been shown in **Figure 2** to visualize the different characteristic temperatures. As per the earlier reports, the interesting properties of relaxor ferroelectric are basically due to the presence of unique polar structure in nanometer size (polar nanoregions: PNRs) along with their response towards the external stimuli [11]. Therefore, it is necessary to understand the origin of PNRs and their effects on the physical properties, as discussed below.

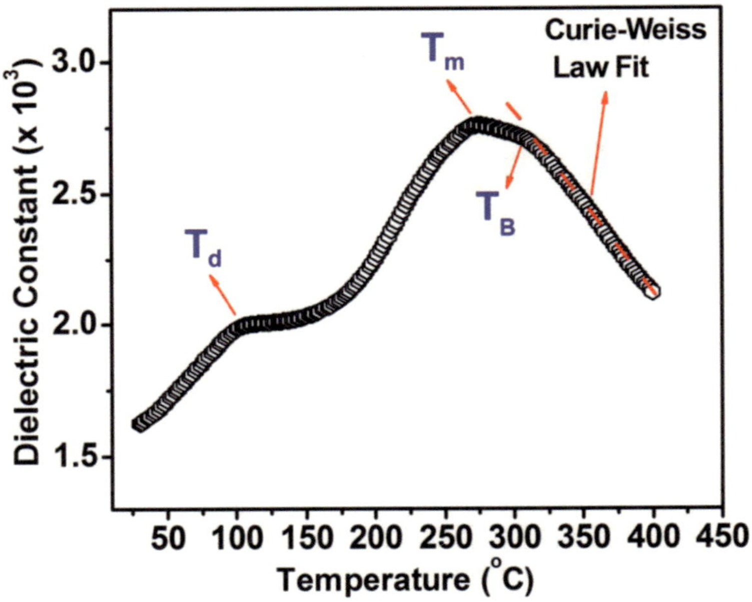

Figure 2.
Various characteristic temperatures are associated with a temperature-dependent dielectric constant of relaxor ferroelectric. Adapted from Ref. [11] and reproduced with permission (self-citation © 2021 IOP).

2.1 Polar nanoregions (PNRs)

The concept of polar nanoregions in relaxor ferroelectric is quite impressive in terms of fundamental understanding of various peculiar physical properties. Also, it has been reported that the types as well as size of the ferroelectric domains greatly influence the external factors (i.e., electric field, temperature, stress, etc.) and, consequently, affect the different physical properties [12]. Initially, several models have been proposed to describe dielectric anomalies in relaxor ferroelectrics, which lead to formulate the concept of the origin of dynamic and formation of polar nanoregions (PNRs) [10]. The conclusions of various models have been described here to correlate the abnormal behavior of relaxor ferroelectric through the presence of nanosize polar domains. Smolenskii proposed the presence of compositional fluctuation in nanometer-scale within the crystal structure by considering a statistical distribution for the phase transition temperature [13].

Furthermore, Cross extended the Smolenskii theory to superparaelectric model, which is associated with the relaxor ferroelectric behavior with thermally activated superparaelectric clusters [9]. Viehland *et al.* have reported the presence of cooperative interaction between superparaelectric clusters and forms the glass-like freezing behavior similar to the spin-glass system [14]. In later study, Qian and Bursil reported the role of random electric field on the formation and dynamic of polar clusters originated from the nanoscale chemical inhomogeneity and defects [15]. According to their theory, the dispersive nature of relaxor ferroelectric is associated with the variation of cluster size and correlation length (the distance above which the polar clusters become non-interactive) as a function of temperature. Although various models and theories have been explained the relaxor ferroelectric behavior, more additional theories and experiments need for its detailed description. Therefore, the dynamic of polar nanoregions (PNRs) with respect to an electric field as well as the temperature has been mentioned here to understand the clear picture (indirectly) of PNRs within relaxor ferroelectric in terms of dielectric and ferroelectric properties.

In the normal unpoled ferroelectric, the large-size domains with specific dipole moments are randomly distributed throughout the material. However, the domains are switched and aligned in the field direction as far as possible with the application of the threshold electric field (coercive field). This process leads to the formation of macroscopic polarization. The polarization is further enhanced with the electric field and reached to the maximum polarization (P_{max}). Usually, the normal ferroelectric exhibits large values of characteristic parameters (P_{sat}, P_r, E_c, *and* hysteresis loss) [16]. However, it is interestingly observed that the ions (cations) in parent compounds replaced by the small number of foreign ions (in terms of different ionic charge and radius) lead to break the long range ferroelectric polar order domains and, converted into a large number of polar islands which are known as polar nanoregions (PNRs) [17, 18]. In the present context, the fundamental factor related to the formation of polar nanoregions has been related to the appearance of intrinsic inhomogeneity in the material due to the compositional fluctuation at the crystallographic sites and structural modification of the unit cell. N. Qu *et al.* reported the evolution of ferroelectric domains with PNRs in the ferroelectric ceramic (refer **Figure 1** [19]). It is clearly observed that the ferroelectric hysteresis loop becomes slim (having low hysteresis loss) in nature with the increase of the percentage of nanodomains. It occurs due to the reduction of correlation length between order parameters (dipole moments). As compared to the normal ferroelectrics, it is difficult to realize the P_{max} at the moderated electric field. It is mainly due to the absence of inducing the long-range polar ordering as well as an increase in the local random field.

Furthermore, the temperature-dependent dielectric constant ($\varepsilon' \sim T$) of relaxor ferroelectric provides detailed information about the dynamic and formation of polar nanoregions (PNRs). The temperature evolution of PNRs in ($Bi_{0.5}Na_{0.5}TiO_3$)-$SrTiO_3$-$BaTiO_3$ based relaxor ferroelectric has been explained along with supported by the impedance ($Z''/Z''_{max} \sim f$) as well as electric modulus ($M''/M''_{max} \sim f$) analysis (refer Figure 11 [20]). In general, the relaxor ferroelectric exhibits the paraelectric phase at high temperature with zero net dipole moment (randomly oriented dipole moments) and well follows the Curie–Weiss law. However, the nucleation of polar nanoregions (PNRs) initiates during the cooling at the particular temperature known as Burn's temperature (T_B). This is the temperature limit below which Curie–Weiss law deviates from $\varepsilon' \sim T$ curve as shown in **Figure 4** with T_B = 286 °C [20]. Also, T_B represents the manifestation of phase transition from paraelectric to ergodic relaxor with the formation of PNRs. This observation is well supported by the occurrence of normalized impedance and electric modulus around the same frequency above the Burn's temperature. This happens due to the higher flipping frequency of the local dipoles at the high-temperature region. Further reduction of temperature (< T_B) leads to enhance formation of polar nanoregions (PNRs) and, subsequently, maximized in volume near dielectric maxima temperature (T_M) with broad dispersive peak, which is supported by the observation of separated normalized impedance and electric modulus peaks. The separation in peak position between Z''/Z''_{max} Vs M''/M''_{max} indirectly confirmed the maximum number of PNRs with the distribution of localized dipolar relaxation. At this temperature ($T_m \sim 150$ °C), the polar nanoregions associate with maximum number relaxation times as indicated by the broad dielectric maxima. Also, the dipole moments of PNRs are randomly distributed with approximately zero remanent polarization. The small polarized entities extend throughout the grains with different correlation lengths. Its dynamic behavior with respect to external stimuli has been observed successfully through various experiments such as in situ transmission electron microscope, neutron diffusion scattering, and piezoelectric force microscopy, and so on [21, 22]. Furthermore, the interaction between the polar nanoregions increases continuously with the reduction of temperature (< $T_m \sim 150$ °C), which leads to initiate the freezing process of PNRs with the phase transition from ergodic to nonergodic state (ferroelectric). After that, the re-oriental flipping of polar nanoregions get a freeze at a particular temperature, known as freezing temperature (T_f) and, represented by the nonergodic state (as shown in **Figure 4** with $T_f \sim 76$ °C).

2.2 Theory of relaxor ferroelectric

As per the literature, the formation of PNRs is related to compositional fluctuation at the lattice sites of the crystal structure. To understand the fundamentals behind the formation of PNRs in the relaxor ferroelectric, ABO_3 type general perovskite compound with different charge states in A & B-site (i.e. $A'_xA''_{1-x}B'_yB''_{1-y}O_3$) can be considered. The randomness of A (A', A") and B (B', B") sites depends upon the ionic sizes, a charge of cations, and distribution of cations in the sub-lattice. Depending upon the randomness, the distribution of domains is classified into two categories such as LRO (long-range order) and SRO (short-range order). SRO can be represented by the continuous with order distribution of cations on the neighboring sites, and the size of the order domains are extended in the range from 20 to 800 Å in diameter, whereas LRO is extended above 1000 Å diameter [23]. Hence, the diffuse phase transition behavior is the characteristic feature of a disordered system in which the random lattice disorder-induced the dipole impurities and defects at the crystal sites. The correlation length (r_c) in a highly ordered ferroelectric compound is defined as the extended length up to which dipole entities are correlated with each other. It is basically larger than the lattice constant (a) in normal ferroelectric and strongly dependent upon the temperature. With reducing the temperature, the correlation length increases gradually and promotes the growth of polar order in a corporative manner with long-range order in FE (ferroelectric) at $T < T_c$ [23]. However, the correlation length (r_c) in relaxor ferroelectric significantly reduced, which forms the polar nanoregions by frustrating the long-range ordering in FE similar to the dipolar glass-like behavior.

According to Landau-Devonshire theory of free energy (F) of a uniaxial ferroelectric material can be expressed in terms of electric field and polarization by ignoring the stress field as follow; [24]

$$F = \frac{1}{2}aP^2 + \frac{1}{4}bP^4 + \frac{1}{6}cP^6 + \ldots - EP \tag{1}$$

Where E is the electric field, P is electrical polarization, a, b, and c are the unknown coefficient with temperature-dependent. Here, the even powers of P are considered because the energy should be the same for $\pm P_s$. P_s is the saturation polarization. The equilibrium configuration can be estimated by finding the minima value of F [24]. The phenomenological description of ferroelectrics states that the spontaneous polarization varies continuously with temperature and attains zero value at the Curie's temperature (T_c: dielectric maxima temperature) in normal ferroelectric. However, the properties of relaxor ferroelectrics, which depend upon polarization in the order of P^2 have been measured experimentally at above the maximum dielectric temperature (shown in **Figure 3**), which supported the presence of polar nanoregions (PNRs) [25, 26]. Furthermore, different formulations have been extensively used to estimate the various characteristic parameters and the degree of relaxor ferroelectric behavior (diffuseness) of materials as described below.

The relaxor ferroelectrics exhibit the broad diffuse maxima in the temperature dependence dielectric permittivity. For a typical RFE, strong dispersion behavior of

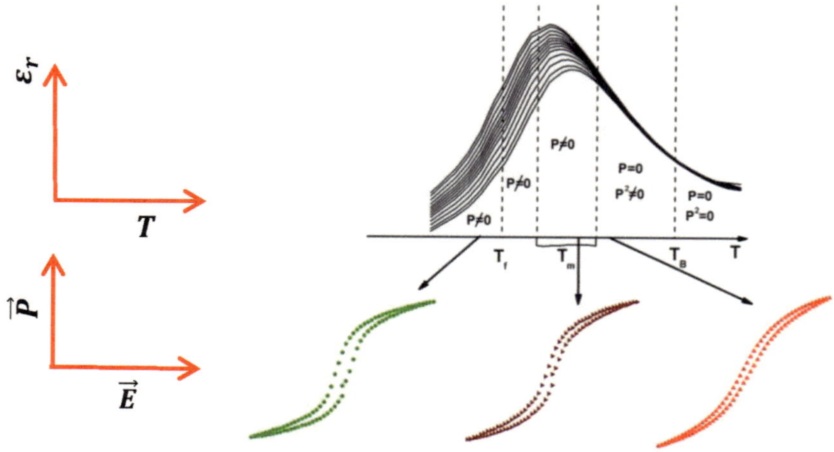

Figure 3.
Confirmation of the presence of polar nanoregions at above the maximum dielectric temperature (T_m), ε_r', T, $\vec{P}$ and E are the permittivity, temperature, electrical polarization, and electric field, respectively. Adapted from ref. [23] (open access).

dielectric permittivity $\varepsilon'(T,\omega)$ with applied frequency observes at the lower side of dielectric maxima temperature (T_m) and becomes independent at the higher temperature region. Also, the T_m shifts towards the higher temperature with an increase in frequency, which is clearly shown in **Figure 3**. Also, the reduction in relaxation processes below T_m in RFEs suggests the onset of relaxor freezing. Therefore, the characteristic relaxation time diverges near freezing temperature (T_f) as per the following Volgel-Fulcher (VF) law relation [27, 28].

$$f = f_0 \exp\left(\frac{-E_a}{K_B\left(T_m - T_f\right)}\right) \tag{2}$$

Here, f is the measuring frequency, and E_a, T_f an f_0 are the parameters. In general, T_f is believed to be the temperature corresponds to freezing of the dynamic of PNRs due to the increase in the cooperative interaction between them. A typical VF relation for $BiAlO_3$-(x) $BaTiO_3$ solid solution with x = 0.05 and 0.1 has been shown in **Figure 4** [29].

Furthermore, the Curie–Weiss (C-W) law is well known to describe the ferroelectric to paraelectric phase transition for normal ferroelectric and, well fitted above the phase transition temperature (Curie temperature: T_c). However, it deviates for relaxor ferroelectrics due to the presence of diffuse phase transition. Therefore, reciprocal of *C-W* law can be used to estimate the degree of deviation (diffuseness) for RFEs by the following relation [30].

$$1/\varepsilon_r' = \frac{T - T_c}{C} \tag{3}$$

Here, ε_r' is the dielectric permittivity, C is the Curie temperature. It has been explained that, the C-W law is well fitted at above the Burn's temperature (T_B). Hence, the degree of diffuseness (ΔT) in term of temperature can be calculated as follow. [31, 32]

$$\Delta T = T_B - T_m \tag{4}$$

To visualize the above relations for the experimental observations, the fitted curve has been shown in **Figure 5** [11]. The separation between maximum dielectric temperature and Burn temperature represents the characteristic of diffuse phase transition.

Furthermore, Uchino and Nomura quantified the dielectric material's relaxor behavior from the ferroelectric to the paraelectric phase transition. They have estimated the diffuse behavior by employing the modified Curie–Weiss (MCW) law. According to this law, the reciprocal of the permittivity can be expressed as; [33].

$$\frac{1}{\epsilon'} - \frac{1}{(\epsilon_{\max})'} = C(T - T_m)^{\gamma}\ 1 \le \gamma \le 2 \tag{5}$$

Here γ and C are the constants. $\gamma = 1$ for the normal ferroelectric material. $1 \le \gamma \le 2$ represents the relaxor ferroelectric behavior and, $\gamma = 2$ is known as the completely diffuse phase. The MCW law is well fitted for $(La_x(Bi_{0.5}Na_{0.5})_{1-1.5x})_{0.97}Ba_{0.03}TiO_3$ based relaxor ferroelectric with different mole fraction of La (refer **Figure 3** [34]).

According to Smolenskii's work, the temperature dependence of dielectric permittivity above T_m is extensively studied to explain the DPT (diffuse phase transition) of relaxor ferroelectrics. Besides this, there are other models reported to describe DPT behavior quantitatively at above the maximum dielectric temperature ($T > T_m$). One of the most applied models is based on Lorentz-type empirical relation as follow [35, 36].

$$\frac{\varepsilon_A}{\varepsilon_r} = 1 + \frac{(T - T_A)^2}{2\delta_A^2} \tag{6}$$

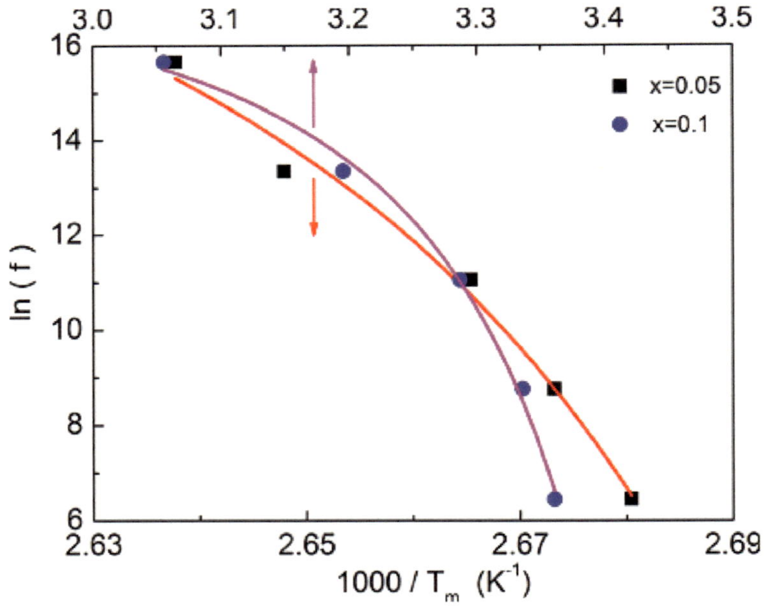

Figure 4.
Volgel-Fulcher relation plot for $BiAiO_3$-(x) $BaTiO_3$ solid solutions with x = 0.05 & 0. Adapted from ref. [29] and reproduced with permission (© 2021 AIP).

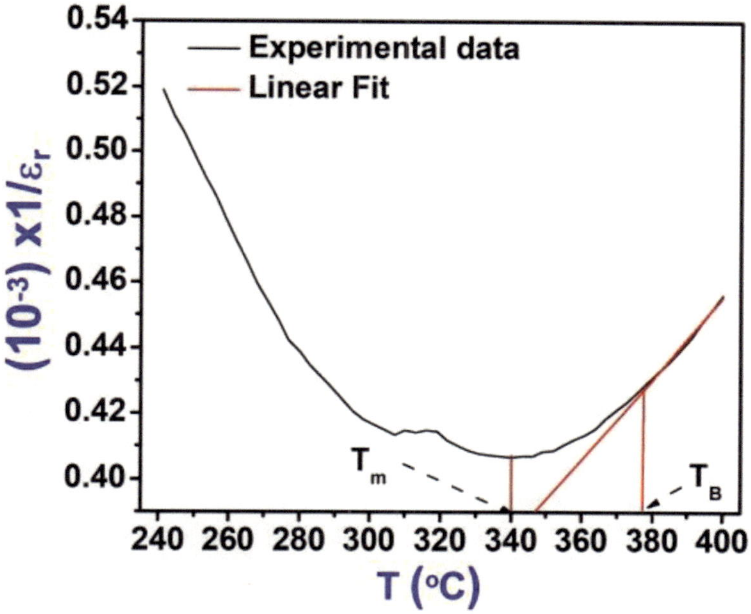

Figure 5.
Variation of $\frac{1}{\varepsilon_r}$ *as a function of temperature, and the solid straight lines represent the C-W law at a particular frequency.* T_B *is the burn temperature. Adapted from ref. [11] and reproduced with permission (self-citation © 2021 IOP).*

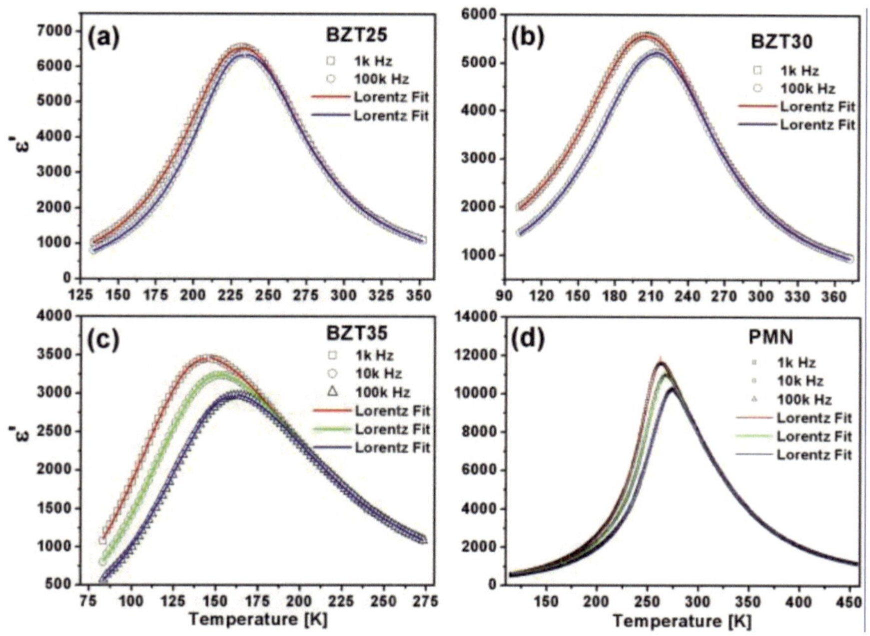

Figure 6.
The temperature-dependent dielectric permittivity of (a) BZT25 (b) BZT30, (c) BZT35 and (d) PMN ceramic fitted with Lorentz relation (solid lines). Adapted from ref. [37] and reproduced with permission (© 2021 AIP).

Where T_A ($T_A \neq T_m$) and ε_A are the fitting parameters. T_A can be defined as the temperature corresponding to the dielectric peak and ε_A is the dielectric constant at T_A. δ_A is the diffuseness parameters which is independent of temperature and frequency. Hence, the DPT behavior of RFEs represents in terms of temperature (T_m-T_A) and dielectric constant ($\varepsilon_r - \varepsilon_A$) . Lei *et al.* have reported that the Lorentz formula is well fitted in both lower and higher temperature regions in $\varepsilon_r \sim T$ curve for Ba ($Ti_{0.8}Sn_{0.2}$)O_3 relaxor ferroelectric [37]. Similarly, Lorentz type relationship in temperature-dependent dielectric permittivity in Ba (Zr_xTi_{1-x})O_3 solid solutions, $PbMg_{1/3}Nb_{2/3}O_3$ relaxor with diffuse phase transition has been reported by S. Ke *et al.* A typical plot of Lorentz formula is shown in **Figure 6**.

3. Materials aspects

Currently, there are several classes of relaxor ferroelectrics available based on different compositions. However, the existing relaxor ferroelectrics are further modified with suitable foreign elements to tune the physical properties. On the basis of material aspect, the relaxor ferroelectrics are classified into two categories as lead-based and lead-free relaxor ferroelectrics. In the present scenario, the lead-based RFEs are dominated extensively on the commercial market due to their superior dielectric, ferroelectric and piezoelectric properties. However, the restriction of lead-based compounds due to their environmental issue (toxic in nature) leads to an increase in the demand for lead-free relaxor ferroelectrics. Therefore, lead-based and lead-free relaxor ferroelectrics are briefly discussed [38, 39]. In addition, the concept of morphotropic phase boundary has been discussed to understand the formation of various solid solutions.

3.1 Concept of morphotropic phase boundary

This term is very common in the field of complex ferroelectric solid solutions. Basically, the term 'morphotropic' referred to the crystal phase transition due to the change in the composition and which is different from the polymorphic phase boundary (PPB) [40]. PPB is generally referred to as a change in phase transition with respect to the temperature. Morphotropic phase boundary represents the phase transition between rhombohedral and tetragonal ferroelectric phases with respect to the variation of composition or pressure. In the vicinity of MPB, the materials exhibit maximum dielectric and piezoelectric properties due to the presence of structural inhomogeneity [41]. The most common and well-known ferroelectric material, i.e., Lead Zirconate Titanate [$Pb(Zr_{0.52}Ti_{0.48})TiO_3$/PZT] lies near morphotropic phase boundary (MPB), as shown in **Figure 7**. Basically, PZT solid solution is the competing of coexistence phases, i.e., $PbZrO_3$ with rhombohedral symmetry (*R3c*) and $PbTiO_3$ with tetragonal symmetry (*P4mm*) as shown in **Figure 7**. However, the origin of extraordinary physical properties near MPB is still a matter of debate. In recent years, researchers have been drawn attention towards preparing solid solutions near MPB of different ferroelectric oxides [41]. The MPB composition of PZT is associated with the 14 possible polarization axis (i.e., 6 from tetragonal and 8 from the rhombohedral), which leads to reduce the energy barrier (i.e., Landau free energy) of the crystal system and, subsequently enhanced the piezoelectric, dielectric, and ferroelectric properties. The dielectric, polarizability, and electromechanical coefficients can be further increased by modifying MPB composition with different suitable acceptors or donor dopants [41].

3.2 Lead-based relaxor ferroelectrics

The most widely used ferroelectric materials for the different applications are derived from the solid solution of (x) $PbTiO_3$-(1-x) $PbZrO_3$. The lead-based

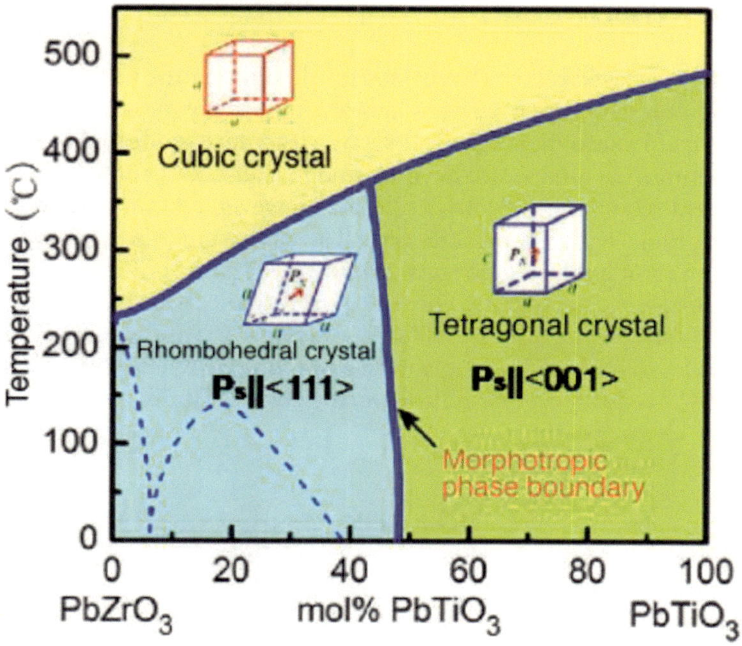

Figure 7.
Lead Zirconate Titanate (PZT) solid solution near morphotropic phase boundary (MPB) composition between $PbZrO_3$ with rhombohedral symmetry (R3c) and $PbTiO_3$ tetragonal symmetry (P4mm). Adapted from ref. [41] (open access).

solid solutions exhibit different characteristics properties ranging from normal ferroelectric-relaxor- antiferroelectric. Therefore, the lead-based relaxor ferroelectrics are quite important for the modern device industries. To date, the PZT is considered a vastly used piezoelectric material. The relaxor properties in PZT have been successfully developed by incorporating foreign elements in A/B sites with isovalent and aliovalent dopants (donor and acceptor dopants). Some of the dopants which have been used in the PZT crystal system are as follows: Isovalent types (Ba^{2+}, Sr^{2+} for Pb^{2+} site & Sn^{4+} for Zr^{4+} and Ti^{4+} sites), Donor types (La^{3+}, Nd^{3+}, Sb^{3+} for Pb^{2+} sites & Nb^{5+}, Ta^{5+}, Sb^{5+}, W^{6+} for Zr^{4+} and Ti^{4+} sites) and acceptor types (K^+, Na^+ for Pb^{2+} & Fe^{3+}, Al^{3+}, Sc^{3+}, In^{3+}, Cr^{3+} for Zr^{4+} and Ti^{4+} sites). Other than PZT, the lead-based relaxor ferroelectrics are $Pb(Mg_{1/3}Nb_{2/3})O_3$(PMN) or $Pb(Sc_{1/2}Ta_{1/2})O_3$(PST), $Pb_{1-x}La_x(Zr_{1-y}Ti_y)_{1-x/4}O_3$(PLZT),$Pb(Zn_{1/3}Nb_{2/3})O_3$(PZN) $Pb(Mg_{1/3}Ta_{2/3})O_3$(PMT), $Pb(Sc_{1/2}Nb_{1/2})O_3$(PSN),$Pb(In_{1/2}Nb_{1/2})O_3$(PIN), $Pb(Fe_{1/2}Nb_{1/2})O_3$(PFN), $Pb(Fe_{2/3}W_{1/3})O_3$(PFW) and the solid solutions:$(1-x)Pb(Mg_{1/3}Nb_{2/3})O_3-xPbTiO_3$(PMN-PT) and $(1-x)Pb(Zn_{1/3}Nb_{2/3})O_3-xPbTiO_3$ (PZN-PT) [42–45]. Therefore, few recent experimental results of lead based ferroelectrics have been explained here to understand the formation and dynamic of PNRs. As per the literature, (1-x) $[Pb(Mg_{1/3}Nb_{2/3})O_3]$-(x) $[PbTiO_3]$ solid solution exhibit normal ferroelectric properties near MPB (x = 0.30 to 0.35) [46]. Also, the structural fluctuation from rhombohedral to tetragonal through intermediate phases (i.e. monoclinic/orthorhombic/triclinic) has been observed with respect to x vary from 0.13 to 0.30. The relaxor properties in above solid solution have been developed (refer **Figures 5** and 7 [46]) with substitution of optimum mole fraction (4%) of Sr^{2+} in place of Pb^{2+}. The formation of compositional fluctuation leads to form short-range order PNRs and, subsequently, reduces the remanent polarization and coercive field and the appearance of diffuse phase transition. The presence of PNRs significantly enhanced the dispersive nature of dielectric permittivity with frequency [46].

3.3 Lead-free relaxor ferroelectrics

Although lead-based relaxor ferroelectrics are dominating in the electronic markets, lead-free ceramics have been focused intensively for last few years due to the restriction of hazardous substances such as lead, lead oxide and heavy metals. There is no equivalent alternative as compared to lead-based compounds, particularly PZT based relaxor ferroelectrics till now. However, certain lead-free relaxor ferroelectric groups with a perovskite crystal structure are impressed by the current researchers for their enhanced physical properties in terms of dielectric, ferroelectric, and piezoelectric properties. Those relaxor ferroelectrics are typical classified as follows: (a) barium titanate ($BaTiO_3$/BTO) based-, (b) potassium sodium niobate ($K_{0.5}Na_{0.5}NbO_3$/KNN) based-, (c) bismuth sodium titanate ($Bi_{0.5}Na_{0.5}TiO_3$/BNT) based- and (d) bismuth layer structured ferroelectrics (BLSFs) [47].

3.3.1 BTO based relaxor ferroelectrics

At room temperature, BTO exhibits stable electrical properties (dielectric and ferroelectric), good electrochemical coupling ($k_{33} \sim 0.50$), high-quality factor, low dielectric loss, but limited by low T_c (120 °C–135 °C) and d_{33} (~190 pC/N). Also, it follows the subsequent structural phase transitions from cubic (>120 °C–135 °C)-tetragonal (120 °C to 20 °C)-orthorhombic (20 °C to −80 °C)-rhombohedral (<−80 °C). In general, $BaTiO_3$ exhibits normal ferroelectric and follows the Curie–Weiss law at ferroelectric to the paraelectric phase transition. The $BaTiO_3$-$BaSnO_3$ solid solution was the first BTO based compound in which relaxor ferroelectric behavior was observed. After that, the relaxor behavior in BTO has been developed by designing the A and B-sites with incorporation of both heterovalent and isovalent ionic substitutions. Currently, there are several modified BTO ceramics available with diffuse phase transition. The available BTO based relaxor ferroelectric systems are $BaTiO_3$-$CaTiO_3$, $BaTiO_3$-$BaZrO_3$-$CaTiO_3$ [$d_{33} \sim$ 620pC/N for $Ba_{0.85}Ca_{0.15}Ti_{0.90}Zr_{0.10}O_3$], $BaTiO_3$-$BiFeO_3$-$Bi(Mg_{0.5}Ti_{0.5})O_3$, $BaTi_{0.8}Sn_{0.2}O_3$, $Ba(Ti_{0.94}Sn_{0.03}Zr_{0.03})O_3$ $BaTiO_3$–$La(Mg_{0.5}Ti_{0.5})O_3$, $BiTiO_3$-(x)$Bi(Mg_{2/3}Nb_{1/3})O_3$ and so on [47–49].

3.3.2 KNN based ceramic system

The KNN system is one of the most promising lead-free alternatives due to its high T_c (~410 °C), high P_r (~33μcm^{-2}), and large K_p (~0.454). Basically, KNN is the solid solution of two perovskite compounds, i.e., $KNbO_3$ (orthorhombic: ferroelectric) and $NaNbO_3$ (orthorhombic: antiferroelectric). In general, the KNN forms the morphotropic phase boundary as similar to PZT [47]. It exhibits moderate dielectric, ferroelectric and piezoelectric properties as compared to PZT. Similar to BTO, the relaxor behavior of KNN has been developed by introducing other elements through interrupting the long rang polar ordering and, forms the PNRs as evidenced by several experimental results. Some of the KNN based relaxor ferroelectrics with physical properties are $(K_{0.48}Na_{0.535})_{0.942}Li_{0.058}NbO_3$ [$d_{33} \sim 314$, $K_{33} \sim 41$, $T_c \sim$ 490 °C], $(K_{0.44}Na_{0.52}Li_{0.04})(Na_{0.86}Ta_{0.10}Sb_{0.04})O_3$ [$d_{33} \sim$ 416pC/N], $0.96(K_{0.5}Na_{0.5})_{0.95}Li_{0.05}Nb_{1-x}Sb_xO_3$–$0.04BaZrO_3$, 0.5wt%Mn-KNN ($T_c \sim$ 416 °C, $d_{33} \sim$ 350 pCN^{-1}), $(Na_{0.44}K_{0.515}Li_{0.045})Nb_{0.915}Sb_{0.045}Ta_{0.05}O_3$($d_{33} \sim$ 390pC/N, $T_c \sim$ 320 °C $K_{33} \sim 0.49$), $0.96(K_{0.4}Na_{0.6})(Nb_{0.96}Sb_{0.04})O_3$–$0.04Bi_{0.5}K_{0.5}Zr_{0.9}Sn_{0.1}O_3$ ($d_{33} \sim$ 460pC/N, $T_c \sim$ 250 °C, $K_{33} \sim$ 0.47), $(Na_{0.5}K_{0.5})_{0.975}Li_{0.025}Nb_{0.76}Sb_{0.06}Ta_{0.18}O_3$ ($d_{33} \sim 352$, $T_c \sim$ 200 °C, $K_{33} \sim 0.47$) [48, 50].

3.3.3 BNT based ceramic systems

BNT is one of the promising lead-free materials to compete with PZT for actuator applications. It exhibits relaxor ferroelectric properties with relatively large remanent polarizations ($P_r \sim 38\ \mu C/cm^2$), large coercivity ($E_c \sim 73$ kV/cm), and high Curie temperature (~320 °C). It follows the character of an ergodic relaxor with room temperature rhombohedral crystal symmetry. The physical properties of several BNT based binary and solid ternary solutions near MPB composition along with the substitution of various cations have been reported; such as BNT-$ATiO_3$ (A = Ca^{2+}, Sr^{2+}, Ba^{2+} and Pb^{2+}), BNT-$KNbO_3$, BNT-$Bi_{0.5}Li_{0.5}TiO_3$, BNT-$Bi_{0.5}K_{0.5}TiO_3$(BNT-BKT), BNT-$K_{0.5}Na_{0.5}NbO_3$(BNT-KNN), BNT-BKT-KNN, BNT-BT-KNN, BNT-BKT-$BiFeO_3$, BNT-BKT-$BaTiO_3$-$SrTiO_3$ and so on [50, 51].

3.3.4 Bismuth layer structured ferroelectrics (BLSFs)

Recently, BLSFs are considered as lead-free relaxor ferroelectrics due to its excellent fatigue properties. The general formula of BLSFs is $(Bi_2O_3)^{2+}$ $(A_{m-1}B_mO_{3m+1})^{2-}$ with A-site occupy by mono-, di- or trivalent ions, B-site occupies by tetra-, penta- or hexavalent ions with appropriate size [47]. The possible A-site elements could be K^+, Na^+, Ca^+, $Sr.^+$, Pb^{2+}, Ba^{2+}, La^{3+}, Bi^{3+}, Ce^{3+}, etc. and B-site could be Ti^{4+}, Nb^{5+}, Ta^{5+}, W^{6+}, Mo^{6+}, etc. The number "m" (=1, 2, 3, 4, and 5) is the number of BO_6 octahedra in the $(A_{m-1}B_mO_{3m+1})^{2-}$ perovskite blocks. BLSFs exhibit high T_c, low tanδ, low ε_r, and decent aging resistance. The above mentioned various classes of relaxor ferroelectrics exhibit different unique relaxor behavior depending upon the formation of PNRs due to the compositional fluctuation in the crystallographic sites.

4. Possible applications

The relaxor ferroelectrics can have wide range of technological applications due to its intriguing physical properties in terms of dielectric, ferroelectric and piezoelectric. As per the earlier discussion, the fundamental origin of unusual behavior in RFEs is mainly due to the presence of polar nanoregions (PNRs). In this section, few important applications of RFEs in modern technologies based on the specific physical property have been briefly discussed for the scientific community. In addition, the systematic approaches have been formulated for individual properties.

i. Currently, the electrical energy storage systems (EESSs) with high energy density and power density are the essential components for the various types of electronics. Out of several EESSs, dielectric capacitors (DCs) are widely used for delivering energy due to its high power density (PD). Power density (P) is defined as the amount of energy delivered by the device per unit time per unit volume. It can be defined as; [52]

$$P = \frac{W_e}{tV} \tag{7}$$

Where, W_e is the energy storage by the device, t and V are the time and volume, respectively. To obtain a high power density, it is essential to increase the energy storage density of the device. The energy storage density of DCs directly related to the dielectric displacement (D) and external applied electric field (E) by the following relation [52].

$$W_e = \frac{1}{2}\varepsilon_r E_b^2 \tag{8}$$

Where ε_r is the relative dielectric permittivity of the material and E_b is the electric field corresponding to the breakdown strength (BDS). However, the high electric field (high BDS) requires high energy storage in the linear dielectric materials. Hence, it creates the hindrance for portable and compact device applications. On the other hand, ferroelectric materials are the possible option to use for energy storage devices at the moderate electric field. Different characteristic parameters related to the energy storage capacity of a nonlinear dielectric material (ferroelectric) are mentioned here [53].

$$\text{Total energy storage density}: W_{store} = \int_0^{P_{max}} E\,dP \tag{9}$$

$$\text{Total recovery energy density (useful energy)}: W_e = \int_{P_r}^{P_{max}} E\,dP \tag{10}$$

$$\text{Efficiency of useful energy storage density}: \eta(\%) = \frac{W_e}{W_{store}} = \frac{W_e}{W_e + W_{loss}} \times 100 \tag{11}$$

Where P_{max} and P_r are the maximum polarization and remanent polarization of the ferroelectric hysteresis loop, respectively. E is the applied electric field. W_{loss} is the total loss of energy storage density. Therefore, the ferroelectric materials' energy storage density can be enhanced by increasing the difference between maximum

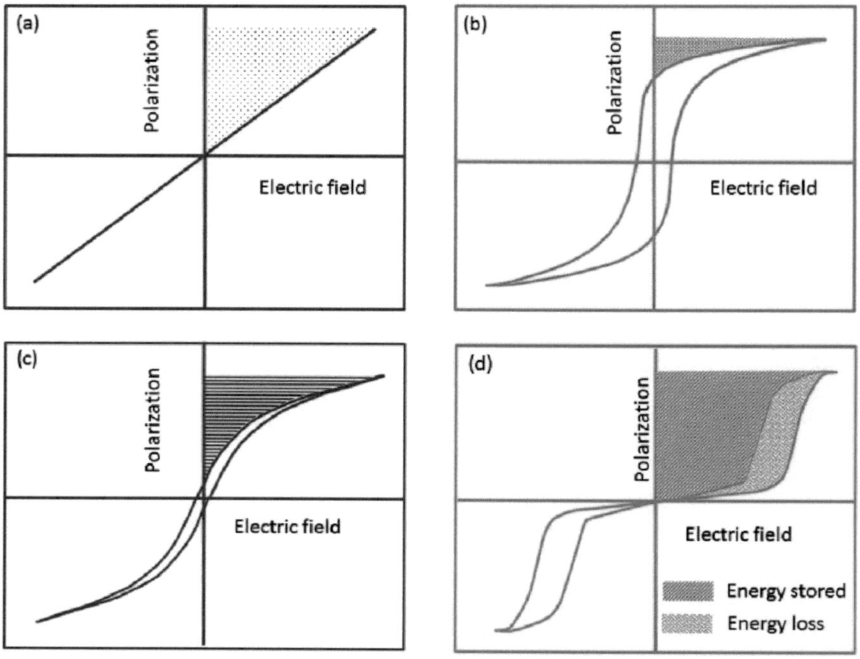

Figure 8.
Typical polarization versus electric field with energy storage capacity and energy loss of different dielectric materials (a) linear, (b) ferroelectric, (c) relaxor ferroelectric, and (d) antiferroelectric. Adapted from ref. [54] (open access).

polarization (P_{max}) and remanent polarization (P_r). A brief description of various dielectric materials' energy storage density is shown pictorially in **Figure 8** [54].

Out of the different dielectric materials, antiferroelectrics exhibit the highest energy storage density. However, there are certain limitations of antiferroelectric which hindrance for its technological applications, such as; few availability (mostly lead-based) and less life cycles (i.e., degradation of antiferroelectric behavior with time). Hence, relaxor ferroelectrics become important for the dielectric capacitor with high energy storage density along with power density due to its slim/constricted ferroelectric hysteresis loop.

ii. Nowadays, refrigerators are widely used as an essential requirement in various sectors starting from household to industrial applications. Basically, the refrigerator is the process to keep cooling a space or substance at below room temperature. The most popular refrigerator process is based on the vapor-compression of the refrigerant [55]. The most common refrigerant used for cooling purposes is chlorofluorocarbons (CFC), which is toxic in nature and cause the Ozone layer depletion. Therefore, several alternative methods have been developed for the cooling system. Out of that, electrocaloric effect based refrigerator becomes one of the emerging cooling technologies that evolved recently. The Electrocaloric (EC) effect observes in the materials having dipolar entities (electric dipoles) and depends upon the interaction between an electric field with the order parameter (dipole moments), which leads to varying the randomness (entropy) of the dipoles in the system [55]. The change in entropy of the material leads to heating or cooling of the respective system under adiabatic conditions. One can refer to the details on electrocaloric effect, including theory, measurement, and application, as reported by Z. Kutnjak *et al.* [56]. Therefore, the EC effect is directly related to the degree of disorder in the materials. As per the material prospective, relaxor ferroelectrics exhibit a higher EC effect than normal ferroelectric due to the presence of a larger number of short-range order polar islands, i.e., polar nanoregions (PNRs).

There are different ways to estimate the entropy change in the materials and subsequently quantified the electrocaloric effect for technological applications. Among them, the indirect method has been widely employed for different systems by the scientific community to calculate the entropy (ΔS) as well as temperature change (ΔT) of the material as mentioned below [56].

$$\Delta S = -\frac{1}{\rho}\int_{E_1}^{E_2}\left(\frac{\partial P}{\partial T}\right)_E dE \tag{12}$$

$$\Delta T = -\frac{1}{\rho}\int_{E_1}^{E_2}\frac{T}{C}\left(\frac{\partial P}{\partial T}\right)_E dE \tag{13}$$

Here, ρ is the density of the sample. The pyroelectric ratio $\left(\partial P/\partial T\right)_E$ can be estimated from the polarization-temperature (*P–T*) curve. *C* is the specific heat of the material, E_1 and E_2 are the initial and final applied electric fields, respectively. ΔT is the change in temperature with the applied electric field. Some of relaxor ferroelectrics in bulk as well as thin film form are $Pb_{0.92}La_{0.08}Zr_{0.65}Ti_{0.35}O_3$ thin film (ΔT = 3.5 K at 700 kV/cm), [57] $[Bi_{0.5}(Na_{0.72}K_{0.18}Li_{0.1})_{0.5}]_{1-x}Sr_xTiO_3$ (ΔT = 2.51 K at 65 kV/cm), [58] $Ba_{0.85}Ca_{0.15}Zr_{0.10}Ti_{0.90}O_3$ (ΔT = 1.479 K at 60 kV/cm), [59] $0.65(0.94Na_{0.5}Bi_{0.5}TiO_3–0.06BaTiO_3) –0.35SrTiO_3$ thin film ($\Delta T \sim$ 12 K at 2738 kV/cm) [60] and so on.

iii. Recently, the electric field-induced strain (electromechanical property) in ferroelectric ceramics has focused intensively due to their wide range of applications in sensors, actuators, MEMS, medical ultrasonic imaging, and so on. In a nutshell, the degree of induced strain in a material can be represented by the parameter S_{max}/E_{max}. Hence, there are two ways to increase the overall strain performance (d_{33}: piezoelectric strain coefficient), i.e. either increase the maximum strain (S_{max}) or reducing the driving electric field (E_{max}). It is well known that, PZT exhibits outstanding electromechanical properties due to the structural instability between rhombohedral and tetragonal crystal symmetries near morphotropic phase boundary [61]. In general, the non-ergodic relaxor phase shows the irreversibly high piezoelectric coefficient (d_{33}) with minimum electrostrain due to the inverse piezoelectric effect. However, compositional fluctuation induced non-ergodic to ergodic relaxor exhibit large repeatable strain due to the phase transition from ergodic relaxor to electric field induce intermediate/metastable ferroelectric phase. The main difference between ergodic and non-ergodic relaxor states is the presence of PNRs. The evolution of PNRs size with respect to the applied electric field plays an important role on the recoverability of the electric field-induced phase transition. Therefore, the ergodicity of relaxor ferroelectric can be enhanced by substituting heterovalent ions that alter the local random electric field and subsequently increase the overall electrostrain. B. Gao *et al.* reported the unexpectedly high piezoelectric response in Sm doped PZT54/46 with $d_{33} \sim 590$ pC/N, kp ~ 57.1% and $S_{max} \sim 0.31\%$ [62].

Some of the reported relaxor ferroelectrics are BNT-BKT-Bi$(Ni_{2/3}Nb_{1/3})O_3$ solid solution ($S_{uni} \sim 0.51\%$ at 65 kV/cm, $d_{33} \sim 890$ pm/V at 45 kV/cm),[63] 0.66PNN-0.34PT ($d_{33} \sim 560$ pC/N),[64] 0.97[0.94$Bi_{0.5}Na_{0.5}TiO_3$–0.06$BaTiO_3$]-0.03$AgNbO_3$ ($d_{33} \sim 721$ pm/V at 60 kV/cm),[65] (BNKT–BST–La0.020 ($S_{max} \sim 0.39\%$, $d_{33} \sim 650$ pm/V) [66] and so on. In addition, novel semiconductor-relaxor ferroelectric based 0–3 type composite has been reported for high temperature piezoelectric applications. Those are ZnO (semiconductor)-(BNT-BTO) (relaxor ferroelectric), ZnO- (BZT-BCT), etc. [67].

iv. Due the multifunctionality of ferroelectrics, it has wide range of applications in modern technologies. Out of that, voltage control dielectric permittivity behavior of ferroelectric is widely used for microwave devices such as resonators, phase shifters, filters and so on [68]. For these applications, the material's tunability (change in capacitance with applied DC bias) should be high with a minimum loss factor. For normal ferroelectrics, tunability attends maximum value near transition temperature i.e. Curie's temperature and hence, different materials can be used for different applications [69]. For example, $SrTiO_3$ like materials are useful at cryogenic temperature, whereas $BaTiO_3$ and PZT are most suitable for room temperature. As per the fundamental aspect, the tunability of a material directly depends on how fast the dipoles respond to the external electric field. The following relations can be used to estimate the tunability quantitatively [69].

$$Tunability(\%) = \frac{\varepsilon(E_0) - \varepsilon(E)}{\varepsilon(E_0)} \times 100\% \quad (14)$$

Where, $E_0 = 0$ kV/cm, and E is the electric field at which the tunability to be measured. In addition to high tunability, high $Q = 1/tan\delta$ or low loss requires to reduce the power loss. Overall, the figure of merit (FOM) for a dielectric tunability material is an important parameter to select the material for applications [69].

$$FOM(K\ factor) = tunability \times Q = \frac{\varepsilon(E_0) - \varepsilon(E)}{\varepsilon(E_0)} \times 1/tan\delta \qquad (15)$$

Interestingly, it has been observed that the tunability in relaxor ferroelectric exhibits a higher value as compared to normal ferroelectrics. It happens due to the presence of short-range ordering PNRs, which respond very quickly with the applied electric field compared to the long-range ordering of dipoles in normal ferroelectrics. Maiti *et al.* reported the lead free relaxor ferroelectric $Ba(Zr_{0.35}Ti_{0.65})O_3$ with room temperature tunability ~44% and K-factor ~ 234 at 40 kV/cm [69]. Similarly, Z. Liu *et al* reported the tunable dielectric properties of $K_{0.5}Na_{0.5}NbO_3$-(x) $SrTiO_3$ (x = 0.16, 0.17, 0.18, 0.19) relaxor ferroelectric with highest tunability ~31.6% for x = 0.16 [70]. There are other relaxor ferroelectrics available in literature having tunable dielectric properties.

5. Future aspect

The above discussion provides the possible applications of relaxor ferroelectrics in modern technologies. Besides the mentioned applications, the RFEs can also use in the field of multiferroics, pyroelectric, photoferroelectric, and so on. Hence, the requirement of relaxor ferroelectrics in the current ceramic market will increase significantly in the near future. Therefore, it is essential to focus on the research and development of RFEs in both academic and industry to find novel materials.

6. Conclusion

The present chapter describes the fundamental understanding of normal ferroelectrics to relaxor ferroelectrics and its possible applications in the modern technologies. The intriguing properties of relaxor ferroelectrics originate due to the presence of polar nano regions as a result of compositional fluctuation at the crystal structure. Although there are several theories available to explain the origin and dynamic of PNRs with external stimulus, it needs to establish the proper structure-properties relation for relaxor ferroelectrics. As per the material aspect, lead-based materials are dominating in the current markets. Therefore, the design of new lead-free relaxor ferroelectrics with enhanced physical properties is required for future applications.

Acknowledgements

Ths work was supported by Indian Institute of Technology Patna (IIT Patna).

Conflict of interest

The authors declare no conflict of interest.

Appendices and nomenclature

MPB	Morphotropic phase boundary
PNRs	Polar nanoregions
REFs	Relaxor ferroelectrics
MCW law	Modified Curie–Weiss law
BTO	$BaTiO_3$ (Barium Titanate)
BNT	$Bi_{0.5}Na_{0.5}TiO_3$ (Bismuth Sodium Titanate)
KNN	$K_{0.5}Na_{0.5}TiO_3$ (Potassium Sodium Niobate)
BKT	$Bi_{0.5}K_{0.5}TiO_3$ (Bismuth Potassium Titanate)
PZT	$PbZr_{0.48}Ti_{0.52}O_3$ (Lead Zirconate Titanate)

Author details

Lagen Kumar Pradhan and Manoranjan Kar*
Department of Physics, Indian Institution of Technology Patna, Patna, Bihar, India

*Address all correspondence to: mano@iitp.ac.in

Refrences

[1] Cross L E, Newnham R E, "History of ferroelectrics" American Ceramic Society (Ceramics and Civilizaion, Volume II: High-Technology Ceramics-Past, Present, and Future), 1987.

[2] Johnsson M and Lemmens P, Wiley Online Library, 2007. DOI: org/10.1002/9780470022184

[3] Ramesh R, Dutta B, Ravi T S, Lee J, Sands T, and Keramidas V G, Appl. Phys. Lett. 1994; 64: 1588

[4] Shkuratov S I, World Scientific, 2019, 456p. DOI:org/10.1142/10958.

[5] Mudinepalli V R, Leng F, Ceramics 2019; 2: 13-24.

[6] Nguyena M D, Nguyend C T Q, Vu H N, Rijnders G, Current Applied Physics 19; 2019: 1040-1045

[7] Jullian Ch, Li JF, Viehland D, J. Appl. Phys. 2005; 95(8): 4316-4318.

[8] Peláiz B A, García Z O, Calderón F, López N R, Fuentes B J, Phys. Sta. Sol. (b) (2005; 242(9):1864-1867

[9] Cross L E, Ferroelectrics 1987; 76: 241.

[10] Chang W A, *et al.*, Journal of the Korean Physical Society 2016; 68: 12, 1481-1494.

[11] Pradhan L K, Pandey R, Kar M, J. Phys. Condensed matter 2020; 32: 045404.

[12] Alikin D, Turygin A, Kholkin A. Shur V, Ceramics Materials 2017; 10(1): 47.

[13] Smolienskii G A, J. Phys. Soc. Japan 1970; 28: 26-30.

[14] Viehland D, Jang S J, Cross L E, J. Appl. Phys. 1990; 68: 2916-2921.

[15] Quian H, Bursill L A, Int. J. Mod Phys. B 1996; 10: 2007-2025.

[16] Jin L, Li F, Zhang S J, J. Am. Ceram. Soc. 2014; 97: 1-27.

[17] Li F, Zhang S J, Damjanovic D, Chen L Q, Shrout T R, Adv. Funct. Mater. 2018; 28: 1801504.

[18] Wu J Y, Mahajan A, Riekehr L, Zhang H F, Yang B, Meng N, Zhang Z, Yan H X, Nano Energy 2018; 50: 723-732.

[19] Qu N, Du H, Hao X, J. Mater. Chem. C 2019; 7: 7993.

[20] Praharaj S, Rout D, Anwar S, Subramanian V, Journal of Alloys and Compounds 2017; 706: 502-510.

[21] Yoshida M, Mori S, Yazmamoto N, Uesu Y, Kiat J M, Ferroelectrics 1998; 217: 327-333.

[22] Caranoni C, Menguy N, Hilczer B, Glinchuk M, Stephanovich V, Ferroelectrics 2000; 240:241-248.

[23] Kumar A, Correa M, Ortega N, Kumari S, Katiyar R S, IntechOpen 2013. DOI: 10.5772/54298

[24] Rabe K M, Ahn C H, Triscone J M, Physics of Ferroelectrics: A Modern Perspective, springer 2007.

[25] Burns G, Dacol F H, Solid State Commun. 1983; 48: 853.

[26] Westphal V, Kleemann W, Glinchuk M, Phys. Rev. Lett. 1992; 68: 847.

[27] Vogel H, Physik Z. 1921; 22: 645.

[28] Fulcher G, J. Am. Ceram. Soc. 1925; 8: 339.

[29] Zheng S, Odendo E, Liu L, Shi D, Huang Y, Fan L, Chen J, Fang L,

Elouadi B, Journal of Applied Physics 2013; 113: 094102.

[30] Cross L E, ferroelectric 1994; 151: 305-320.

[31] Raddaoui Z, Kossi S E, Dhahri J, Abdelmoulab N, Taibi K, RSC Adv. 2019; 9: 2412.

[32] Macutkevic J, Banys, J. Bussmann H A, Bishop A R, Physical Review B 2011; 83: 184301

[33] Uchino K, Nomura S, Ferroelectr. Lett. 1982: 44: 55.

[34] Yu Z, Chen X, Physica B 2016; 503:7-10.

[35] Bokov A A, Ye Z G, Solid State Commun. 2000; 116: 105.

[36] Bokov A A, Bing, Y H, Chen W, Ye Z G, Bogatina S A, Raevski I P, Raevskaya S I, Sahkar E V, Phys. Rev. B 2003; 68: 052102,

[37] Shanming K, Huiqing F, Haitao H, H. Chan L W, Applied Physics Letters 2008; 93: 112906.

[38] Damjanovic D. Lead-Based Piezoelectric Materials: Piezoelectric and Acoustic Materials for Transducer Applications. Springer (Berlin); 2008. 59-79p. DOI: org/10.1007/978-0-387-76540-2_4

[39] Takenaka T, Nagata H, J. Eur. Ceram. Soc. 2005; 25: 2693-2700.

[40] Abdessalem M. B, Aydi S, Aydi A, Abdelmoula N, Sassi Z, Khemakhem H, Applied Physics A 2017; 123: 583.

[41] Aksel E, Jones J L, Sensors 2010; 10: 1935-1954.

[42] Kang M G, Jung W S, Kang C Y, Yoon S J, Actuators 2016; 5: 5.

[43] Liou Y C, Wu L, Huang J L, Yu C H, Japanese J. Appl. Phys. 2002; 41: 1, 3A.

[44] Duran, P, Lozano J F F, Capel F, Moure C, J. Mater. Sci. 1989; 24: 447-452.

[45] Kour P, Pradhan S K, Kumar P, Sinha S K, Kar M, Appl. Phys. A 2016; 122: 591.

[46] Augustinea P, Muralidhar M, Samantad S, Naik S P K, Sethupathid K, Murakamic M, Rao M S R, Ceramics International 2020; 46: 5658-5664.

[47] Wei H, *et al.* J. Mater. Chem. C 2018; 6: 12446.

[48] Shrout T R, Zhang S J, J. Electroceram 2007; 19: 113-126.

[49] Wang T, Jin L, Li C, Hu Q, Wei X, J. Am. Ceram. Soc. 2015; 98 [2]: 559-566.

[50] Seshadri R, Hill N A, Chem. Mater 2001; 13: 2892-2899.

[51] Reichmann K, Feteira A, Li M, Materials 2015; 8: 8467-8495.

[52] Veerapandiyan V, Benes F, Gindel T, Deluca M, A Review, Materials 2020; 13: 5742.

[53] Jan Abdullah, Liu H, Hao H, Yao Z, Cao M, Arbab S A, Tahir M, Appiah M, Ullah A, Emmanuel M, Ullah A, Manan A, J. Mater. Chem. C 2020; 8: 8962.

[54] Chauhan A, Patel S, Vaish R, Bowen C R, 2015; 8: 8009-8031.

[55] Valant M, Progress in Material Science 2012; 57: 980-1009.

[56] Kutnjak Z, Rozic B, Pirc R, Electrocaloric Effect: Theory, Measurements, and Applications .Wiley online library; 2015. 10.1002/047134608X.W8244

[57] Nguyen M D, Houwman E P, Rijnders G, J. Phys. Chem. C 2018; 122:15171-15179.

[58] Zhang L, Zhao C, Zheng T, Wu J, ACS Appl. Mater. Interfaces 2020; 12: 33934-33940

[59] Hanani Z, *et al.*, RSC Adv. 2020; 10: 30746.

[60] Yang C, Han Y, Feng C, Lin X, Huang S, Cheng, X, Cheng Z, ACS Appl. Mater. Interfaces 2020;12: 6082-6089.

[61] Singh A K, Pandey D, Yoon S, Baik S, Shin N, Appl. Phys. Lett. 2007; 91: 192904.

[62] Gao B, Yao Z, Lai, D, Guo Q, Pan W, Hao H, Cao M, Liu H, Journal of Alloys and Compounds 2020; 836:155474

[63] Dong G, Fan H, Liu, L, Ren P, Cheng Z, Zhang S, Current Science 2020; 118: 10

[64] Zhang Y, Liu H, Sun S, Deng S, Chen J, J Am Ceram Soc. 2021;104: 604-612.

[65] Rena, P, Liua Z, Liuc H, Sunc S, Wana Y, Longd C, Shie J, Chenc J, Zhao G, Journal of the European Ceramic Society 2019;39: 994-1001.

[66] Aman U, Kim W, Lee D S, Jeong S J, Ahn C W, J. Am. Ceram. Soc., 2014; 97 [8]: 2471-2478

[67] Zhang Ji, *et al.*, Nature Communications 2015; 6:6615.

[68] Farnell G W, Cermak I A, Silvester P, Wong S K, IEEE Trans. on Sonics and Ultrasonics, 1970;17: 188.

[69] Maiti T, Guo R, Bhalla A S, Applied Physics Letters 2007; 90: 182901.

[70] Liu Z, Fan H, Zhao Y, Dong G, J. Am. Ceram. Soc. 2016; 99 [1]: 146-151.

Chapter 5

Piezoelectricity and Its Applications

B. Chandra Sekhar, B. Dhanalakshmi, B. Srinivasa Rao, S. Ramesh, K. Venkata Prasad, P.S.V. Subba Rao and B. Parvatheeswara Rao

Abstract

The piezoelectric effect is extensively encountered in nature and many synthetic materials. Piezoelectric materials are capable of transforming mechanical strain and vibration energy into electrical energy. This property allows opportunities for implementing renewable and sustainable energy through power harvesting and self-sustained smart sensing in buildings. As the most common construction material, plain cement paste lacks satisfactory piezoelectricity and is not efficient at harvesting the electrical energy from the ambient vibrations of a building system. In recent years, many techniques have been proposed and applied to improve the piezoelectric capacity of cement-based composite, namely admixture incorporation and physical. The successful application of piezoelectric materials for sustainable building development not only relies on understanding the mechanism of the piezoelectric properties of various building components, but also the latest developments and implementations in the building industry. Therefore, this review systematically illustrates research efforts to develop new construction materials with high piezoelectricity and energy storage capacity. In addition, this article discusses the latest techniques for utilizing the piezoelectric materials in energy harvesters, sensors and actuators for various building systems. With advanced methods for improving the cementations piezoelectricity and applying the material piezoelectricity for different building functions, more renewable and sustainable building systems are anticipated.

Keywords: piezoelectric effect, ferroelectricity, actuators, sensors, buzzers

1. Introduction

Technical application of Piezoelectricity phenomenon first discovered by Pierre and Jacques curie and Jacques curie in 1880 [1] and thereafter soon understood from the crystallographic point of view had a very slow start because for decades only a few suitable materials were available. In spite of their small piezoelectric effect, quartz crystals continue to dominate as components for frequency control since the early days of radio engineering [2], this is due to their extremely sharp resonance curves, which are stable with respect to temperature and aging. The first ferroelectric material, Rochelle Salt [3] was found out to the suitable for broadband applications in the year 1920.Stability problems encountered with these crystals, which are produced from aqueous solutions, restrict their application to phonograph pick-ups.

Over the past period the spheres of application of piezoelectric materials in modern techniques have been considerably enlarged. In this relation the requirements to their properties are continuously growing. A great number of the piezoelectric materials have been developed in several countries, yet research in this field is still in active. The efforts of researchers are concentrated on the problem of purposeful development of the materials with desirable combination of their properties. The wide spread application of the piezoelectric effect is based on ferroelectric ceramic materials can be attributed to three main facts:

1. The Piezoelectric effect particularly large in the ferroelectrics.

2. Ceramics can be produced cost effectively. Most of these materials are either impossible or at best very difficult to produce in mono crystalline form.

3. Ceramic materials offer a high degree of variation concerning geometrical shaping on the one hand and physical properties on the other hand by virtue of mixed-crystal formation, creation of differing grain structures, and interaction of various ferroelectric or non- ferroelectric phases.

At present piezoelectric materials based on Barium Titanate ($BaTiO_3$). Lend Zirconate-Lead Titanate (PZT) solid solutions and multi component solid solutions relating to the Perovskite type crystal structure and containing, as a rule, lead titanate or lead zirconate, are mainly used [3].

Most of the improvements in the properties for particular application in the piezoceramics have been achieved either by partially replacing the constituent atoms by other atoms or doping with a small quantity of purity additives. Broadly speaking, all these methods may be considered to the control the ceramic characteristic properties by impurity doping.

Piezoelectricity is the additional creation of an electric charge by the applied stress; this is the direct piezoelectric effect. The charge is proportional to the force, and it is therefore of opposite sign for compression and tension. In terms of dielectric displacement D (charge Q per unit area A) and stress T, it may be written as.

$$D = Q/A = dT \tag{1}$$

There is a converse effect. An applied field E produces a proportional strain S, expansion or contraction depending on polarity.

$$S = dE \tag{2}$$

Therefore, the piezoelectric constant 'd' (Piezoelectric strain coefficient) which is numerically identical for both direct and converse effects.

$$d = D/T = S/E \tag{3}$$

Another frequently used piezoelectric constant is g (piezoelectric voltage coefficient), which give the field produced by a stress and is related to the 'd' constant by the permittivity (ε).

$$g = d/\varepsilon \tag{4}$$

Additional piezoelectric constants which occasionally used are 'e' which relates stress T to field E, and 'h' which relates strain S to field E.

$$T = -eE \tag{5}$$

$$T = -hS \tag{6}$$

Actual definitions are.

$$d = (\partial S/\partial E)_T = (\partial D/\partial T)_E \tag{7}$$

$$g = (-\partial E/\partial T)_{D=} (\partial S/\partial D)_T \tag{8}$$

$$e = (\partial T/\partial E)_{S=} (\partial D/\partial S)_E \tag{9}$$

$$h = (-\partial T/\partial D)_{S=} (-\partial E/\partial S)_D \tag{10}$$

For ceramics and crystals the elastic, dielectric and piezoelectric constants may differ along different axes. For this reason, they are expressed in tensor form.

The hydrostatic strain constant d_h is related to d_{33} and d_{31} as follows:

$$d_h = 2d_{31} + d_{33} \tag{11}$$

$$\text{where } d_{33} = (\partial D_3/\partial T_3)_E = (\partial S_3/\partial E_3)_T$$

$$d_{31} = (\partial D_3/\partial T_1)_E = (\partial S_1/\partial E_3)_T$$

Possibly the best single measurement of the strength of a piezoelectric effect is the electromechanical coupling factor K. When an electric field is applied, it measures the fraction of the electrical energy converted to mechanical energy (or vice versa when a crystal or ceramic is stressed). The actual relationship is in terms of K^2

$$K^2 = \frac{\textit{Electrical energy converted to mechanical energy}}{\textit{Input electrical energy}}$$

$$K^2 = \frac{\textit{Mechanical energy converted to electrical energy}}{\textit{Input mechanical energy}}$$

The piezoelectric, elastic and dielectric constants of poled ceramics are strongly temperature dependent. Heating through the Curie point destroys the effect of poling and causes the piezoelectric properties to disappear [4]. If the sample is heated to just below the Curie point the piezoelectric properties are degraded. A remanant piezoelectric effect produced by polling an initially random orientation ceramic is a strong evidence for ferroelectricity 90^0 walls contribute to the piezoelectric effect since their movement is accompanied by dimensional change and not 180^0 walls because there will be no dimensional change [5]. Ferroelectric materials with high Curie temperature are highly desirable to construct transducers for high temperature piezoelectric applications.

2. Piezoelectricity

Certain crystals become electrically polarized (i.e electric charges appear on their surfaces) when stressed. This phenomenon discovered in 1800 by Pierre and J. Curie is called the piezoelectric effect and the crystals as the piezoelectric crystals Quartz, rochelle salt, tourmaline are the familiar piezoelectric substances.

The inverse effect–that these crystals become strained when polarized has also been observed.

Piezoelectric strains are very small, and the corresponding electric fields are very large. In Quartz for example a field of 1000 V/cm produces a strain of the order of 10^{-7}. Conversely small strains can produce large electric fields.

To understand the origin of the piezoelectric effect, the distribution of the ionic charges of a crystal about their lattice sites. Normally, the distribution is symmetrical, and the internal electric field is zero. But when the crystal is stressed, the charges are displaced. In a piezoelectric crystal this displacement distorts the original charge distribution in such a way that it is no longer symmetrical - for a quartz crystal. A net polarization results in such crystals and when observing the piezoelectric effect. In other crystals, on the other hand, the distribution of charges maintains its symmetry even after the displacement - for a non piezoelectric crystal. Such crystals exhibit no net polarization and hence no piezoelectric effect [6].

It follows that the piezoelectric effect is related to crystal symmetry. The symmetry element involved is essentially the center of inversion. A crystal can exhibit piezoelectric effect only if its unit cell lacks a center of inversion. This is because when there is no center of inversion, only then the charge distribution is distorted so as to produce polarization. However if the center of inversion is present, there is no charge distortion hence no polarization.

It can be proved that of the 32 crystal classes, 21 are non centro symmetrical but since one of these 21 is highly symmetric in other respects, it is piezoelectrically excluded, leaving only 20 piezoelectric classes. However, all crystals belonging to these 20 classes are not observably piezoelectric - in some crystals the piezoelectric effects are too small to be detectable. Thus, the lack of inversion center is a necessary but not sufficient condition to guarantee piezoelectricity.

Piezoelectric effect is extensively used to convert the electric energy into mechanical energy and vice-versa i.e. the piezoelectric substances are used as electromechanical transducers. For instance it is an electric signal that is applied to one end of a quartz rod, the variations in strain generated in the rod in consequence of the effect propagate down the rod constituting what is known as mechanical wave or an acoustic wave. Another important application of piezoelectrics is their use as highly stable oscillators for frequency control [7]. If a quartz crystal is subjected to an alternating voltage at one of its resonant frequencies the crystals will suffer expansion and contraction alternately in consequence of the effect and thus the oscillations of the crystals will be set up. The frequency of these oscillations depends on the dimensions of the specimen and the elastic constants of the material and is stable. Specially cut quartz discs are generally used for this purpose.

Ferro electricity versus piezoelectricity [8]:

1. In piezoelectricity the crystal is polarized by the application of an external stress whereas in ferroelectricity the source of polarization is the dipole interaction energy itself.

2. Both the phenomena occur in noncentrosymmetric crystals, which are 20 in number. Piezoelectricity occurs in all the 20 crystals whereas ferroelectricity only in 10 namely those which provide a favorable axis of polarity.

3. All ferroelectrics are therefore piezoelectric but all piezoelectrics are surely not ferroelectric for example Tourmaline is piezoelectric but not ferroelectric at all.

4. The piezoelectric coefficient is the ratio of the setup charge to the stress applied to a crystallographic axis. The ferroelectrics have very large piezoelectric coefficients.

The phenomenon of piezoelectricity was discovered just over a hundred years ago by the Curie brothers, Pierre and Jaques. The science of Piezoelectricity has proceeded at an uneven face in these one hundred years. Periods of rapid progress have been followed by periods of slow development and sometimes even by periods of no development (Incidentally, this is characteristic of all branches of science). Every time that piezoelectricity has appeared to be exhausted as a science, the discovery of new piezoelectric effects or new piezoelectric materials initiated a new stage of rapid development and opened up new areas for the application of piezoelectricity. Piezoelectricity is currently enjoying a great resurgence in both Fundamental Research and Technical applications.

Piezoelectricity is one of the basic properties of crystals, ceramics polymers and liquid crystals. There are several ways to describe the piezoelectric effect [9]. Perhaps the most common definition is that a material is piezoelectric if the application of an external mechanical stress causes the development of an internal dielectric displacement. This displacement is manifested as an internal electric polarization or a surface electric charge Because of the way in which the elastic stress and dielectric displacement transform during coordinate axis rotation (**Figure 1**) the piezoelectric constants describing the linear relationship form a third order tensor. A simplified mathematical formulation of the piezoelectric effect is given below. More detail treatments of the piezoelectric effect and Converse effect can be found in texts.

It should be noted that the piezoelectric effect is strongly linked to the Crystal symmetry. All crystals are arranged into 32 point groups. Crystals belonging to the 11 centro symmetric point groups cannot show a piezoelectric effect. Crystals belonging to the non centro symmetric point group O also do not exhibit a piezoelectric effect. Nearly all other non metallic crystals belonging to the remaining 20 point groups exhibit a piezoelectric effect of some magnitude, although some of the effects are very small.

The piezoelectric phenomenon can be described as.

$$P_i = P^0{}_i + {}_{jk}Ed_{ijk}\ T_{jk} \tag{12}$$

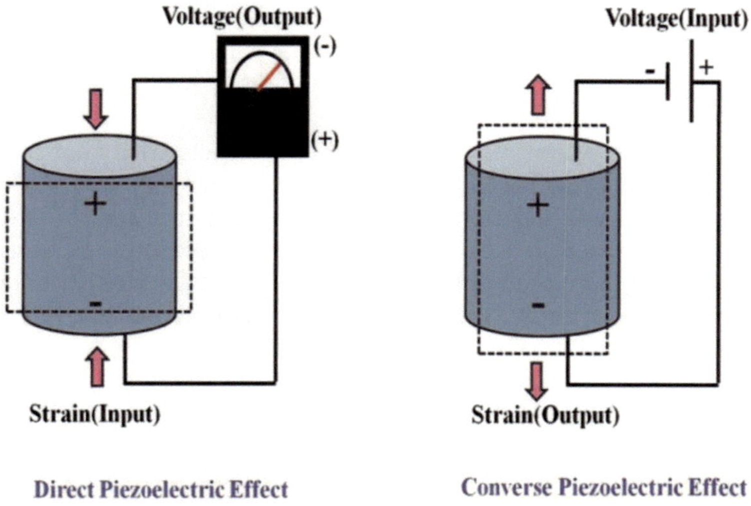

Figure 1.
Piezoelectric effect.

Where P_i is a component of the polarization vector, $P^0{}_i$ the spontaneous polarization and T_{jk} is the stress tensor component. The coefficient d_{ijk} are called the piezoelectric Coefficient and are third rank tensor components.

Piezoelectric materials that are currently receiving much scientific attention include piezoelectric semiconductors, such as gallium arsenide, which have a wide range of interesting properties [10]. An existing goal with these materials is to integrate the piezo device and the semiconductor components on the same substrate. The last decade has witnessed an explosive expansion in research on surface acoustic waves. Most recently, the research has concentrated on layered systems containing piezoelectrics. Another important application of surface acoustic waves has been the development of miniature high-frequency "bulk structure" filters using Lithium niobate and Lithium tantalate crystals for use in consumer electronic applications.

Research into "bulk structure" surface acoustic wave resonators is currently a very active area. Piezoelectric Polymers, thin films and composites are becoming increasingly important. This is evidenced by a series of recent International conferences devoted to PVF_2 and other piezoelectric polymers.

Since piezoelectricity was first discovered the applications of piezoelectric materials have mushroomed. Langevin's work opened the large field of ultrasonics, which now includes detection, nondestructive evaluation, acoustic electricity, acousto optics, and imaging, signal processing, physical acoustics, medical acoustics etc.

Early Works By Cady and Nicolson lead to frequency control including resonators, oscillators and filters [11]. This field initially utilized low frequencies about 100 kHz. As time progressed, higher frequencies were needed and used. The majority of the presently mass produced high frequency piezoelectric filters are based on the Onoe theory of the multimedia resonator. It should be noted that the Onoe theory was inspired by Schockley's theory of energy traps.

Today piezoelectric devices are found in television sets, radios, wristwatches, small computer games, automobiles etc. Many communications and navigation systems used large numbers of very precise piezoelectric resonators for frequency control, generation and selection.

It can be observed that even with quartz the original piezoelectric material, the rate of improvement of the properties of these devices is still in an accelerating phase. For instance, the stability of quartz frequency sources has improved by an order of magnitude every five or six years.

Piezoelectric materials have always played a very significant role in acoustics. In recent times, they have found widespread application as generators, transmitters and detectors of surface acoustic waves.

Of the many biological materials which exhibit piezoelectricity bone belongs to the best investigated ones. Bur has measured various complex piezoelectric constants of bovine bone as a function of frequency, temperature and relative humidity. The presence of water in bone in some piezoelectric constants gives rise to the occurrence of piezoelectric relaxation in others it shifts the relaxation frequency as does the temperature. This piezoelectric relaxation has been qualitatively explained by the two-phase model too. The losses in this case are attributed to a Maxwell Wagner dispersion, which occurs as a result of ionic conduction.

It may appear that the physical mechanism of piezoelectric relaxation by electrical and mechanical interactions between different phases is different from the piezoelectric relaxation as described in the preceding chapters. The basic elements of the piezoelectric relaxation however, are compatible for molecular point defects for two-dimensional defects like domain boundaries and for three-dimensional defects as are the finely dispersed to phase materials. In any case there is a coupling between electrical and mechanical losses, which can be described by the relaxation of defects which are simultaneously as well electric as elastic dipoles. The heterogeneous system entails a higher degree of complexity, example superposition of uncoupled losses, losses by electric conduction, local field effects orientation distributions and others. Therefore the theoretical treatment is clearer in the two-phase model.

Piezoelectric ceramics are prepared for fabricating the electromechanical transducers used in the mechanical frequency filters that find application in long-haul Communications systems. These ceramics have to satisfy specifications that can only be met by utilizing all the possibilities offered by the physical effects of the ferroelectric materials. The required positive temperature coefficient of the frequency constant is realized with the aid of elastic anomalies in the region of ferroelectric phase transitions.

Quartz resonators have been adapted for communications, but in recent years mostly for wrist watches and clocks since the quartz-oscillator circuit which incorporates a piezoelectric quartz crystal resonator has a very stable frequency. Thanks to quartz resonators, time accuracy of wrist watches has been improved rapidly. Quartz resonators for wristwatches and clocks amount to over 60% of total quartz resonators manufactured in Japan. This paper touches upon the characteristics, details of technical advancements, the analysis methods, the manufacturing technique and finally the future trend of quartz resonators for wristwatches.

Recently electronic wrist watches have spread far and wide, small and beautiful ones with high accuracy and many functions in particular. It owes development of various watch parts including the Integrated circuit. Among them the development of the quartz resonator for wristwatches, which produces the time (frequency) standard, is especially splendid.

Elastic vibration of a quartz resonator is transformed into electric Vibration by piezoelectricity because quartz crystal is stable against the ambient temperature, elapsed time and other various environments; frequency of a quartz resonator oscillator is extremely stable. Therefore it has been used in the fields of wireless communications and recently adapted for wrist watches and clocks. It shows the percentage of quartz resonators by fields produced in Japan. As described, quartz resonators for wrist watches and clocks amount to 64.6% of the total number and 40.4% of the gross sales.

3. Applications of piezoelectricity

All the electrical devices nowadays are just not limited to electrical connection in between them but have this piezoelectricity as a common thing in all applications. Cell phones, diesel fuel injectors, grill igniters, ultrasonic transducers, acoustic guitar pickups, vibration sensors, certain printers, and musical greeting cards etc. utilizes piezoelectricity. The additional development of manmade piezo materials which includes piezoelectric ceramics.

The applications of piezoelectricity includes the following fields:

- Piezoelectric Motors

- Actuators in Industrial Sector

- Sensors in Medical Sector

- Actuators in Consumer Electronics (Printers, Speakers)

- Piezoelectricity Buzzers
- Instrument pick-ups
- Microphones
- Piezoelectric Igniters
- Nanopositioning in AFM, STM
- Micro Robotics (Defense)

3.1 Piezoelectric effect works with sensors and motors

To start with, the electric cigarette lighters and gas grills have a high voltage power source when compared with other applications of the piezoelectric effect. In these cases, a hammer strikes a piece of piezo material, which then produces enough current to create a spark that ignites the flammable gas in its presence. However, in other applications like sensors, the hammer is typically replaced by other forms of energy like sound waves - including ultrasound, as hammer is an exciter of the piezo material.

When these are working with sensors, piezo materials will detect even some of the minute disturbances and anomalies, which will make them unique and idealistic devices in industrial nondestructive testing and medical imaging.

In the other perspective, piezoelectric motors can perform highly precise and repeatable movements. This inbuilt makes them excellent devices (**Figure 2**) for the precision movements of sensitive optical devices like telescopes and microscopes.

3.2 Piezoelectric sensors in industrial applications

The industrial sector contributes its applications with piezoelectric sensors for a variety of uses. Some common, everyday uses include:

3.2.1 Engine knock sensors

Engine manufacturers are every now and then facing challenges related to the control of engine devices. Under some non-supporting situations, gasoline engines are susceptible to an undesirable phenomenon known as detonation. When the

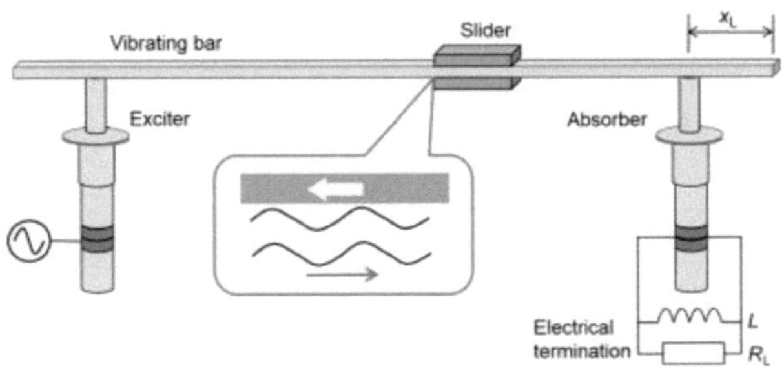

Figure 2.
Traveling wave motor.

DOI: http://dx.doi.org/10.5772/intechopen.96154

process of detonation occurs, the air/fuel charge explodes instead of burning smoothly thereby damaging the engine. Eventually, this is why the designed engines with conservative operational margins at the expense of efficiency — it was to avoid this notorious problem.

With the advancement of the better control systems, the relevant engine parameters may be adjusted in real-time to maximize efficiency and power. If detonation begins to occur, piezoelectric knock sensors can be employed to sense the detonation before it becomes problematic. This gives control systems time to make the required adjustments.

3.2.2 Pressure sensors

In almost all the applications the measurement of dynamic pressure changes, using piezoelectric pressure sensors yields more reliable results than using conventional electromechanical pressure sensors (**Figure 3**). The reason behind this is that piezoelectric devices have a high frequency response and signal conversion without any need of bellows, diaphragm, or any type of mechanical linkage in conjunction with a strain gage or displacement sensor.

3.2.3 Sonar equipment

Sonar Equipment depends especially on piezoelectric sensors to transmit and receive ultrasonic "pings" in the 50-200 kHz range. Along with an ideal frequency response for such applications, piezoelectric transducers have a high power density that enables large amounts of acoustic power to be transmitted from a small package. For instance, a transducer that is only 4″ (100 mm) in diameter may be capable of handling power output greater than 500 watts.

Figure 3.
A piezoelectric pressure sensor.

3.3 Piezoelectric actuators in industrial applications

In this times when piezoelectric sensors are highly valuable to the industrial sector, the industry also makes use of piezoelectric actuators for a variety of applications:

3.3.1 Diesel fuel injectors

In the last decade, regulations on emissions from diesel engines have become increasingly stringent. In addition to this, customers continue to demand quieter engines with improved power and torque curves. In order to meet these stringent demands for compliance and performance, engine manufacturers have resorted to using precisely timed and metered injections of fuel during the combustion process [12].

Besides the working of other applications in piezoelectric devices, a single fuel injector may switch fuel flow with pressures exceeding 26,000 psi (1800 bar) on and off several times in rapid succession during a single power stroke. Such precise control of high-pressure fluid is made possible by using piezoelectric actuators controlling small valves within fuel injectors.

Equipment which involves diesel engine emissions have become increasingly stringent, yet customers' demands for quieter engines with improved power output has led the world's leading firms in fuel injection technologies to invest in extensive research and development. To fulfill the customers demands while remaining compliant and ahead of regulatory pressures, Piezo Diesel Injectors were developed to reduce emissions by making the combustion of fuel within the cylinder more efficient. Such technological progress leads a new phase with possibilities of improved performance and reduced emissions through precisely timed and metered injections of fuel during the diesel combustion process.

3.3.2 Fast response solenoids

Some processes require quick and precise mechanical actuation that is difficult, if it is not satisfied by that process it uses electromagnetic solenoids to achieve. While speed may not always be a concern, power consumption or compactness of size is the most prominent one. In accordance with this, piezoelectric actuators are often able to fill the niche as they provide fast response and low power consumption in small packages, compared to electromagnetic solenoids.

3.3.3 Optical adjustment

Some optics needs to be adjusted or modulated with a wide frequency response and with a minimum number of moving parts. Piezoelectric actuators are often employed in such applications where they provide fast and accurate control over a long service life:

- The angle of a mirror or diffraction grating may need to be precisely varied according to an electrical input. Such applications are often encountered in optical or physics experiments.

- Earth-based telescope arrays are subject to atmospheric distortion, and spacecraft optics are subject to movement and vibration. In such cases, optics may need to be adjusted (shaped or contoured) in real-time by means of a control system. This will compensate for aberrations that would otherwise impede image resolution.

- Some fiber optic converters rely on piezoelectric actuators to modulate the output of a laser.

3.3.4 Ultrasonic cleaning

Piezoelectric Actuators also contribute to ultrasonic cleaning applications. To perform ultrasonic cleaning, objects are immersed in a solvent (water, alcohol, acetone, etc.). A piezoelectric transducer then agitates the solvent. Many objects with inaccessible surfaces can be cleaned using this methodology.

Piezoelectric ultrasonic cleansers (**Figure 4**) provide capabilities such as ultrasonic breaking up of kidney stones and removal of dental plaque. They are used to conduct precise measurements to identify flaws and other anomalies detected between transmitters and receivers of ultrasonic waves.

3.3.5 Piezoelectric motors

One advantage of using piezoelectric materials is that their characteristics are precise and predictable. Thus, expansion and contraction of a piezoelectric actuator can be precisely controlled as long as the supply voltage is controlled. Some motor designs take advantage of this fact by using piezoelectric elements to move a rotor or linear element in precise increments. Precision on the order of nanometers can be achieved with some piezo motor designs. Piezo motors work at a wide range of frequencies but typically work best in a low frequency range.

In addition to their inherent precision, piezoelectric motors can be used in environments with strong magnetic fields or cryogenic temperature — environments where conventional motors are unlikely to work. These unique challenges are present in MRI machines, particle accelerators, and other similar environments.

3.3.6 Stack actuators

Multiple piezoelectric elements will be stacked to replace the displacement achieved for a given voltage. These types of devices are known as stack actuators (**Figure 5**), and they are employed in variety of specialty applications. Compared to

Figure 4.
Piezoelectric ultrasonic cleansers.

conventional electromagnetic actuators, stack actuators have the following unique advantages:

- They can function at cryogenic temperatures or in environments with strong magnetic fields.
- They can produce a large amount of force in a small package
- They can respond almost instantly to input with high rates of acceleration.
- They can achieve extremely high degrees of precision.
- They only consume power when work is actually being performed.

These actuators find their uses in proportioning valves, electrical relays, optical modulation, vibration dampening, and other applications requiring fast or precise control of movement.

3.4 Piezoelectric sensors in medical applications

3.5 Ultrasound imaging

Piezoelectronic Transducers are often used in medical Ultrasound Equipment. Advances in equipment over the decades have enabled improved monitoring of pregnancies and facilitated minimally invasive surgical procedures (**Figure 6**).

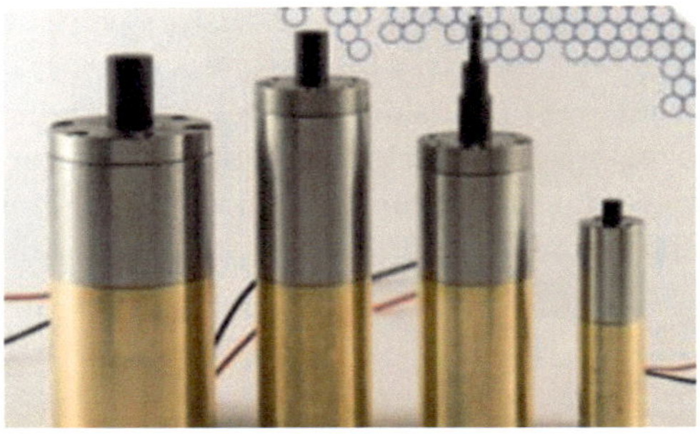

Figure 5.
Piezoelectric stack actuators.

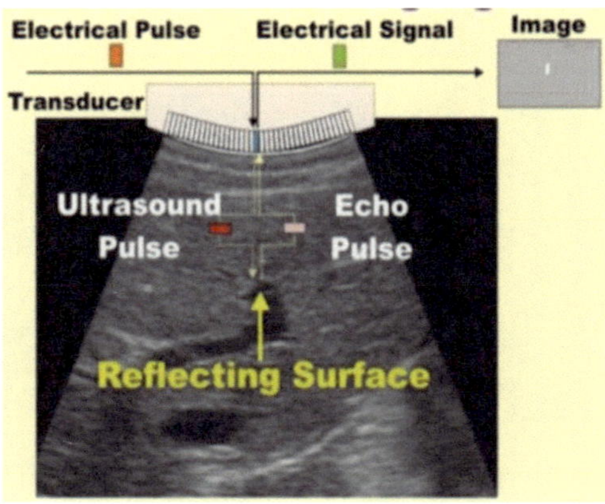

Figure 6.
Ultrasound imaging.

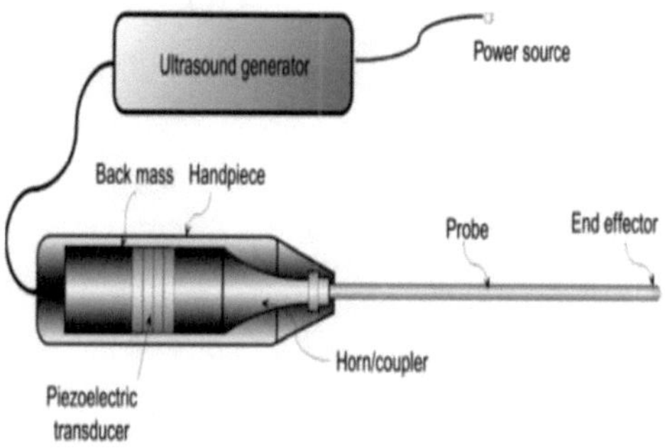

Figure 7.
Ultrasonic procedures.

3.6 Ultrasonic procedures

Some non-invasive medical procedures rely on the use of focused ultrasonic waves to break up kidney stones or destroy malignant tissue (**Figure 7**). Additionally, the advent of the harmonic scalpel has enabled surgeons to simultaneously incise and coagulate tissue during a surgical procedure without the need for cauterization. This leads to less tissue damage, less blood loss, and faster healing times [13].

3.7 Piezoelectric actuators in consumer electronics

Consumer electronics and technology that is sold in stores throughout the country, has piezoelectric actuators used in them.

3.7.1 Piezoelectric printers

There are two main types of printers that use piezoelectric actuators:

3.7.1.1 A dot-matrix printer

In a piezoelectric dot matrix printer (**Figure 8**), piezoelectric actuators in the printer head move needle-like pins that "poke" through a strip of ink tape (similar to a typewriter) against a piece of paper in various patterns to form characters. For most applications, the use of dot-matrix printers has been superseded by other technologies. However, a dot-matrix printer is the only printer technology capable of generating duplicate and triplicate carbon-copy printouts.

3.7.1.2 Inkjet printer

In a piezoelectric inkjet printer, piezoelectric actuators in the printer head act on small diaphragms or otherwise change the geometry of an inkwell so that ink droplets are forced out of an orifice onto paper. This is one of the dominant technologies in the printer market to date (**Figure 9**).

3.7.2 Piezoelectric speakers

Piezoelectric speakers are featured in virtually every application that needs to efficiently produce sound from a small electronic gadget. These types of speakers

Figure 8.
A dot-matrix printer.

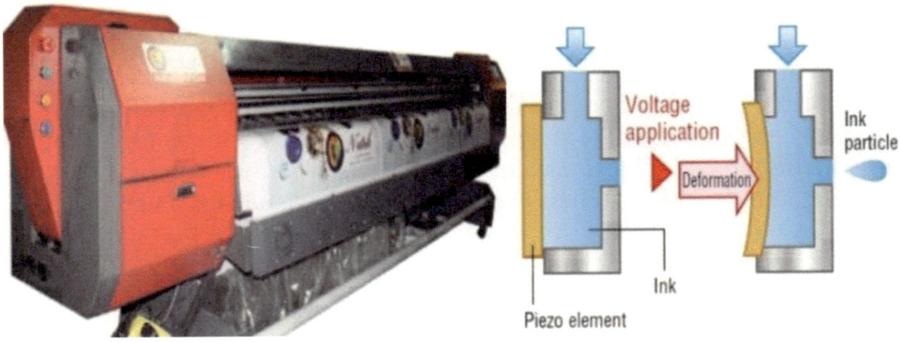

Figure 9.
Inkjet printer working.

Figure 10.
Piezoelectric speakers.

are usually inexpensive and require little power to produce relatively large sound volumes. Thus, piezoelectric speakers (**Figure 10**) are often found in devices such as the following:

- Cell phones
- Earbuds
- Sound-producing toys
- Musical greeting cards
- Musical balloons

3.7.3 Piezoelectric buzzers

Piezoelectric buzzers are similar to piezoelectric speakers, but they are usually designed with lower fidelity to produce a louder volume over a narrower frequency range. Buzzers are used in a seemingly endless array of electronic devices.

3.7.4 Piezoelectric humidifiers

Many cool mist humidifiers use a piezoelectric transducer to transmit ultrasonic sound energy into a pool of water. The ultrasonic vibrations cause fine water droplets to break away and atomize from the surface of the pool where they become entrained in an air stream and enter the desired space.

3.7.5 Electronic toothbrushes

Linear piezoelectric actuators are implemented to vibrate the bristles in some electronic toothbrushes (**Figure 11**).

3.8 Piezoelectricity other applications in daily life

3.8.1 Piezoelectric igniters

This is, perhaps, the most well-known and ubiquitous use of piezoelectricity. In piezoelectric igniters, a button or trigger is used to cock and release a spring-loaded hammer, and the hammer is used to strike a rod shaped piezoelectric ceramic. The sudden mechanical shock to the piezoelectric ceramic produces a rapid rise in voltage that is high enough to jump a sizable spark gap and ignite fuel. Piezoelectric igniters are commonly used for butane lighters, gas grills, gas stoves, blowtorches, and improvised potato cannons.

3.8.2 Electricity generators

Some applications require the harvesting of energy from pressure changes, vibrations, or mechanical impulses. The harvesting of energy is possible by using piezoelectric materials to convert deflections or displacements into electrical energy that can either be used or stored for later use.

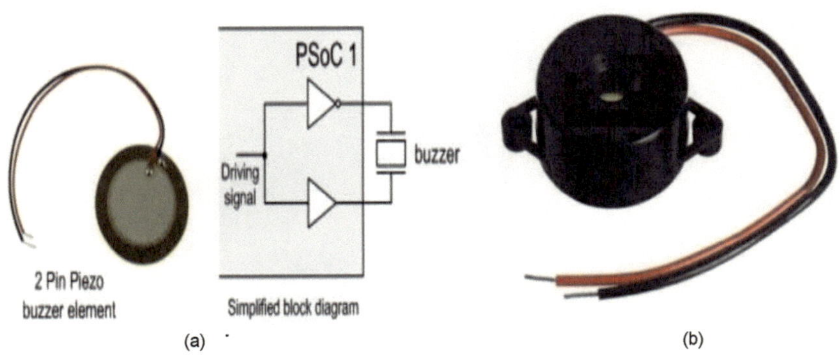

Figure 11.
(a) Piezoelectric buzzer block diagram. (b) Piezoelectric buzzer.

3.8.3 Microelectronic mechanical systems (MEMS)

MEMS devices have become more commonplace as more integrated capabilities are required in smaller packages, such as cell phones, tablet computers, etc. The advantage of MEMS devices is that gyroscopes, accelerometers, and inertial measuring devices can be integrated into chip-sized packages. In order to accomplish such a feat, piezoelectric actuators and sensors are often used.

3.8.4 Tennis racquets

A somewhat unusual application for piezoelectricity integrates piezoelectric fibers into the throat of a tennis racquet along with a microcontroller in the handle. When the tennis player strikes the ball, the racquet frame deflects and generates an electric output that is boosted, reversed, and fed back into the fibers [14]. This is an attempt to cause destructive interference and dampen structural vibration.

3.9 Piezoelectricity in defense applications

The piezoelectricity is used in Defense field for a variety of applications, they are:

3.9.1 Micro robotics

In the field of small robotics, small power-efficient mechanical actuators and sensors are needed. With the use of piezoelectric actuators, building something as small as a robotic fly that can crawl and fly is technically feasible. In fact, a new field of robotic technology known as Micro Air Vehicles aims to build small drones the size of insects or birds that fly using flapping wings (**Figure 12**). They control surfaces just as birds and insects do. These types of feats in miniaturization are possible, in part, by using piezoelectric actuators.

3.9.2 Course-changing bullets

Recently, DARPA invented a.50-caliber bullet that can change course in mid-flight. As absurd as this innovation may sound to some readers, the bullet uses an optical sensor that is mounted on its nose in conjunction with a control system and moveable tail fins to steer itself toward a laser-illuminated target. Although DARPA

Figure 12.
Piezoelectric micro robotics.

has not revealed much about their Extreme Accuracy Tasked Ordnance (EXACTO) bullet, the most likely means of manipulating the tail fins probably involves piezoelectric actuators.

Acknowledgements

I sincerely express my heartfelt gratitude and appreciation to my supervisors Dr. P.S.V.Subba Rao & Dr.B.Parvatheeswara Rao, Department of Physics, Andhra University for their supervision and guidance during the whole duration of my research. I am greatly indebted to my dear parents, family members and my wife Mrs. Niranjani Divya Revu for their constant motivation, inspiration, love and affection, It is my pleasure to acknowledge my colleagues, Finally, I express my sincere thanks to AUTHOURITIES of Vignan Group of Institutions, for his co-operation in extending me the necessary facilities to carry out my research work.

Author details

B. Chandra Sekhar[1*], B. Dhanalakshmi[2], B. Srinivasa Rao[3], S. Ramesh[4], K. Venkata Prasad[5], P.S.V. Subba Rao[6] and B. Parvatheeswara Rao[6]

1 Vignan's Institute of Engineering for Women, Visakhapatnam, India

2 Department of Physics, Vignan's Institute of Information Technology (A), Visakhapatnam, India

3 Welfare Institute of Science, Technology and Management, Visakhapatnam, India

4 School of Technology, GITAM Deemed to be University, Bengaluru, India

5 Department of Physics, Centurion University, Vizainagram, India

6 Department of Physics, Andhra University, Visakhapatnam, India

*Address all correspondence to: cs.beera@gmail.com

References

[1] J. Curie, P. Curie, C. R. Acad. Sci. Paris 91(1880)294. https://doi.org/10.1063/1.2915742

[2] W.G. Cady, "Piezoelectricity", Dover, New Yard, Vol.2 (1964), p.417. https://apps.dtic.mil/dtic/tr/fulltext/u2/a254925.pdf

[3] J. Valasek, Phys. Rev. 15(1920)537. https://doi.org/10.1103/PhysRev.17.475

[4] B. Jaffe, W.R. Cook and H. Jaffe, "Piezoelectric ceramics", Academic Press, New York, 1971. 10.4236/msa.2013.45041

[5] J.M. Harbert, "Ferroelectric transducers and Sensors". Gordon and Breach Science Pub. London, (1982) Chap.1. https://doi.org/10.1007/978-1-4615-3818-9_19

[6] L.G. Van Uitert, S. Singh, H.J. Levinstein, J.E. Geusic and W.A. Bonner, Appl. Phys. Lett. 11(1967) 161. https://doi.org/10.1063/1.1755129

[7] J.E. Geusic, H.J. Levinstein, J.J. Rubin, S. Singh and L.G. Van Uitert, Appl. Phys. Lett. 11(1967)269. https://doi.org/10.1063/1.1755129

[8] S. Singh, J.E. Geusic, H.J. Levinstein, R.G. Smith and L.G. Van Uitert, IEEE J. Quantum Electronics QE-4 (1968)352. https://doi.org/10.1063/1.346406

[9] S.C. Abrahams, H.J. Levinstein and J. M. Reddy, J. Phys. Chem. Solids 27 (1966)1019. https://doi.org/10.1016/0022-3697(66)90074-6

[10] A.W. Warner, G.A. Coqin and J.L. Fink, J. Appl. Phys. 40(1969)4353. https://doi.org/10.1063/1.1657197

[11] T. Yamada, H. Iwasaki and N. Niizeki, J. Appl. Phys. 41(1970)4141. https://doi.org/10.1063/1.1658426

[12] S. Singh, D.A. Draegest and J.E. Geusic, Phys. Rev. B2 (1970)2709. https://doi.org/10.1063/1.4858400

[13] G.W. Taylor, J.J Gagepain, T.R meeker, T.Nakamura and L.A Shuvalov "Piezolelectricity"Vol.4, 1985. https://doi.org/10.1002/crat.2170231222

[14] R.L Singhal, Soid state Physics,1989. https://doi.org/10.1007/978-1-4615-1241-7

Chapter 6

Ferroelectric, Piezoelectric and Dielectric Properties of Novel Polymer Nanocomposites

Maheswar Panda

Abstract

In this chapter, the Ferroelectric, Piezoelectric and Dielectric behavior of novel polymer/ceramic nano-composite (PCC) based on ferroelectric polymer [polyvinyl-edene fluoride (PVDF)] & nano Barium Titanate (*n*-BaTiO3) with different volume fractions of *n*-BaTiO3 (f_{BaTiO3}), prepared through the novel cold pressing method has been discussed. The ferroelectric parameters of PCC are attributed to spherulites of PVDF, the increase of *n*-BaTiO3 and the ordered homogenous structure due to the novel cold pressing. The clustering of ceramic fillers is responsible for randomization of the structures of these composite ferroelectrics for some samples, leading to decrease of electrical polarisations. The piezoelectricity and piezoelectric coefficients of these composites ferroelectrics, increases with increase of ceramic filer content and remains constant beyond a certain ratio. However, the dielectric properties increase linearly as a function of ceramic content due to increase of interfaces/interfacial polarisations. The enhancement of effective dielectric constant (ε_{eff}) is attributed to the large interfacial polarization arising due to the charge storage at the spherulites of PVDF and at the polymer/filler interfaces of PCC and have been explained on the basis of sum effect with the help of the standard models. The achieved lower loss tangent (Tan δ) for the PCC as compared to the polymer/metal composites (PMC) is attributed to the highly insulating nature of PVDF & semiconducting *n*-BaTiO3. The thermal stability of the composites is also maintained due to the higher melting temperature (170°C) of PVDF. The cold pressed PCC based on PVDF are going to act as better polymer ferroelectric/dielectrics for memory and electrical energy storage applications.

Keywords: Polymer ferroelectrics/dielectrics, spherulites, Ferroelectric polymers, Barium titanate, Dielectric constant, Loss Tangent, Polymer nanocomposites

1. Introduction

Polymer ferroelectrics (PF) and polymer dielectrics (PD) are considered recently to be the fascinating materials for their large inherent benefits of non-volatile memory/sensor/piezoelectric/dielectric/pyroelectric/magneto-electric applications [1–17]. The conventional ferroelectric ceramic materials, e.g. BaTiO3, PZT, PbTiO3, etc. being used as memory elements/piezoelectric sensors/actuators/transducers, etc. are suffering from a large number of disadvantages, such as; brittleness, high cost and consume higher energy/longer time for their preparation.

IntechOpen

To overcome these problems, PF are undergoing development based on ferroelectric polymer as well as ferroelectric ceramics. Among the ceramic fillers, BaTiO3 is a very good ferroelectric material and comparably better as others have harmful lead content [16, 17]. Among the various polymers, very few polymers, such as polyvinylidene fluoride (PVDF), polyvinylidene fluoride trifluoroethylene [PVDF Tr(FE)] and Teflon show ferroelectric/piezoelectric behavior and have high dielectric constants [18, 19]. They are also of high breakdown strength, light-weight, flexible and having permanent dipolar polarization. Among the ferroelectric polymers, PVDF shows high piezo-electric coefficients, good ferroelectric behavior. Due to these advantages, PVDF is used as piezo-electric sensors/actuators/memory devices. But the major problem with PVDF is that, the magnitudes of ferroelectric parameters aren't as good as the parameters, obtained from the conventional ceramics, which limit them from direct applications. Similarly the PD are having higher flexibility, non-toxicity, bio-compatibility, low cost, higher visco-elastic properties, etc. [20–31]. Due to their higher energy density/lower loss tangent (Tan δ) with higher breakdown field strength, they are going to be the emerging materials of future for electrostatic energy storage applications. The PD composed of polymers with conductor/ceramic nanoparticles are considered recently to be the demanded materials for electrical energy storage applications [20–36]. For energy storage applications, the maximum stored energy per unit volume is $U = \frac{1}{2} K \varepsilon_0 E_{max}{}^2$, where K is the relative dielectric constant and E_{max} is the maximum electric field, which can be applied to the material (proportional to the breakdown field of the material). Over the last 20 years of research, the ferroelectric polymers e.g PVDF matrix also have been preferred due to it's high static dielectric constant (~15)/higher visco-elastic properties/higher insulating nature as compared to other non-polar polymers. The preferred fillers are high dielectric constant ferroelectric ceramics in development of these PD. However, the development of the polymer-ceramic composites (PCC) have been slowed down, as the effective dielectric constant (ε_{eff}) for these composites were found be very low i.e. $\varepsilon_{eff} \sim 100$ at low frequencies due to the low dielectric constant of the polymers as well as due to the conventional hot molding process conditions. In preparing these PCC, ferroelectric ceramic, such as, PMN-PT, BaTiO3, PbTiO3, etc. with varying particle size are introduced into the PVDF matrix through hot molding and partially the approach becomes effective in order to get better PD [5–10]. For preparing PCC based on PVDF, the traditional mixed technique (solution casting followed by hot molding) is used, during which the spherulites of PVDF get lost [2, 33, 34], which lowers the value of ε_{eff}. Recently Panda et al. [23–30, 32, 33] has shown the importance of spherulites by following the cold pressing technique in preparing the PD based on PVDF, due to which the spherulites of PVDF are retained. The spherulites are responsible in the additional storage of electrical charge due to their additional interfaces, resulting higher interfacial polarization/higher value of ε_{eff} [2, 33, 34].

With the objective of achieving flexibility with low cost/easy processing and higher value of electrical parameters, for device applications, the traditional process condition (hot molding of the thick films prepared from solution casting) is changed to cold pressing developed by our group in which the spherulites of PVDF will be retained for the case of PCC. Hence, PCC of good ferroelectric/piezoelectric/dielectric properties, are developed from good ferroelectric ceramics/good ferroelectric polymers. Since in the PZT/PVDF composites, lead is a toxic component, hence the PCC based on PVDF/n-$BaTiO_3$, with different volume fraction of n-$BaTiO_3$ ($f_{n\text{-}BaTiO3}$), were prepared with the help of cold pressing method. The prepared composites have shown the interesting ferroelectric/piezoelectric/dielectric/conductivity properties and finds suitability for various applications.

2. Experimental details

Polymer composite based on PVDF/*n*-BaTiO3 from 0.0 to 0.60 of volume fraction of nano filler *n*-BaTiO3 (f_{BaTiO3}) were prepared by mechanical hand mixing with Agate mortar/Pestle for 2 hours. The final pellets under room temperature consolidation at a pressure of 30 MPa with the help of a Hydraulic press [2, 33] were prepared. The microstructure investigation on the samples was carried out with the help of FESEM. The ferroelectric hysteresis, i.e. the polarization versus electric field (P ~ E) measurement, is done with the help of a P ~ E hysteresis loop tracer. The piezoelectric coefficient (d_{33}) for the samples is measured with the help of Piezo-electric measurement instrument. The electrical measurements were made on all the PCC in the frequency range of 50 Hz to 5 MHz and in the temperature range of room temperature to 100°C. The dielectric results of the PCC have been understood by fitting with the help of the software Mathematica.

3. Results and discussion

3.1 Microstructure

The FESEM micrographs of pure PVDF are given in **Figure 1a** and **b**. **Figure 1a** and **b** shows the presence of spherulites (the spherical semi-crystalline regions of the polymer). The micrographs of PVDF/n-BaTiO3 composites with different f_{BaTiO3} = 0.2 and f_{BaTiO3} = 0.60 are shown (**Figure 1c–f**). The ordered homogenous structures are also observable and is attributed to the recent novel method of cold pressing as evident from the sample with f_{BaTiO3} = 0.2 (**Figure 1c** and **d**). The spherulites present in the polymer are of diameter of the order of ~0.1 μm (**Figure 1a** and **b**). The *n*-BaTiO3 are also of the order of diameter of the spherulites as they are of size 100 nm, i.e. 0.1 μm, During cold pressing, the *n*-BaTiO3 clusters (**Figure 1c** and **d**) inside the polymer matrix, may have taken the typical shapes. For the sample with f_{BaTiO3} = 0.6 (**Figure 1e** and **f**), shows high level of heterogeneity, as lot of defects and dislocations has emerged in the structure and is responsible for giving a decrease in the ferroelectric & piezoelectric properties of the PF. **Figure 1** reveal slight agglomeration of BaTiO3 nanoparticles in the nanocomposites. The average filler size in the nanocomposites are of ~100 nm. The nano-dispersion of filler in the polymer matrix is well observed. It is also obvious that a variety of interfaces have occurred into the composites, which will be always useful in the storage of electrical charge at the interfaces The large amount of *n*-BaTiO3 into the PCC will also be responsible for giving better ferroelectric/piezoelectric/dielectric properties [2, 34–36].

3.2 Ferroelectric properties

3.2.1 Ferroelectric hysteresis

The ferroelectric properties of the PF, prepared from PVDF/*n*- BaTiO3, the polarization versus electric field (P ~ E) of both the pure materials are given (**Figure 2**) at different voltages. **Figure 2a** shows narrow hysteresis loops for pure PVDF due the mixed phases [α, β, γ, δ & ε] as well as the electrical non-poling of the polymer PVDF.

The polarization versus electric field (P-E hysteresis loop) of the PF under different voltages for various f_{BaTiO3} = 0.2 (**Figure 3a**), f_{BaTiO3} = 0.3 (**Figure 3b**) f_{BaTiO3} = 0.4 (**Figure 3c**) and f_{BaTiO3} = 0.5 (**Figure 3d**) are shown. All the samples show symmetrical P-E hysteresis loops. The loop area increases of with increasing

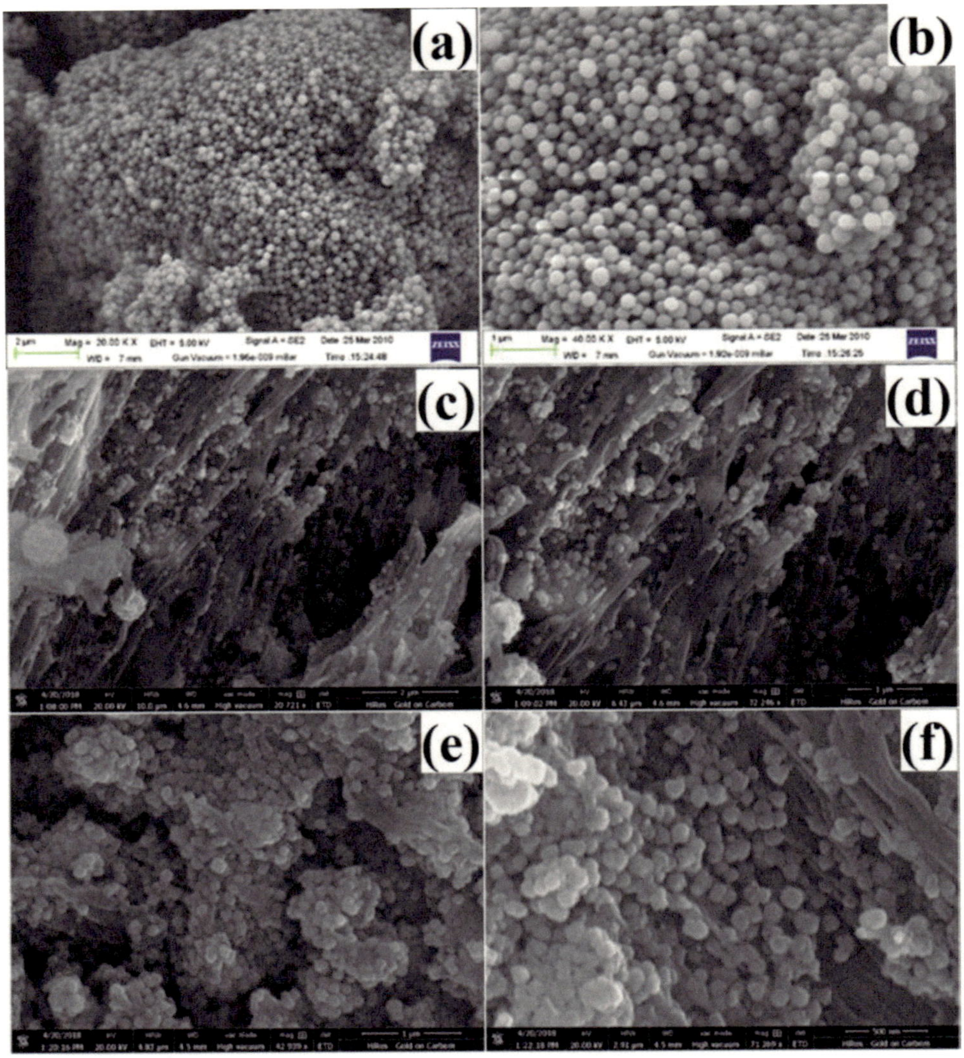

Figure 1.
FESEM micrographs of cold pressed PF (a) pure PVDF (lower resolution) (b) pure PVDF (higher resolution) (c) f_{BaTiO_3} = 0.2 (lower resolution) (d) f_{BaTiO_3} = 0.2 (higher resolution) (e) f_{BaTiO_3} = 0.6 (lower resolution) (f) f_{BaTiO_3} = 0.6 (higher resolution).

the dc voltage from 5 kV to 10 kV. A comparison of P-E hysteresis loops, demonstrate that the samples with f_{BaTiO3} = 0.2 & 0.3 shows better ferroelectric hysteresis (higher hysteresis loop area) as compared to f_{BaTiO3} = 0.4 & 0.5. For precise assessment, the hysteresis loops at 8 kV for all the samples shows that the hysteresis loop area increases as a function of f_{BaTiO3}, up to 0.30 (**Figure 1a–d**). On the other hand, beyond f_{BaTiO3} = 0.30, the heterogeneity/disordered structure is accountable for the decrement of ferroelectric properties and that is also accredited to the clustering of n-BaTiO3 into the polymer medium (**Figure 1e** and **f**). It is also experiential that with rising the field, the saturation polarization (P_s), remnant polarization (P_r) and the coercive field (E_c) also increases as a function of f_{BaTiO3} and the finest result is obtained for the sample with f_{BaTiO3} = 0.3.

To have a thorough analysis and cross examination of the ferroelectric behavior of the PF, the P-E hysteresis loop of all the samples as a function of f_{BaTiO3} for different fields from 5 kV/cm to 8 kV/cm is shown in **Figure 4**. At all the fields, the ferroelectric behavior from the P-E hysteresis loops, are characterized by the change of P_r, P_s and E_c.

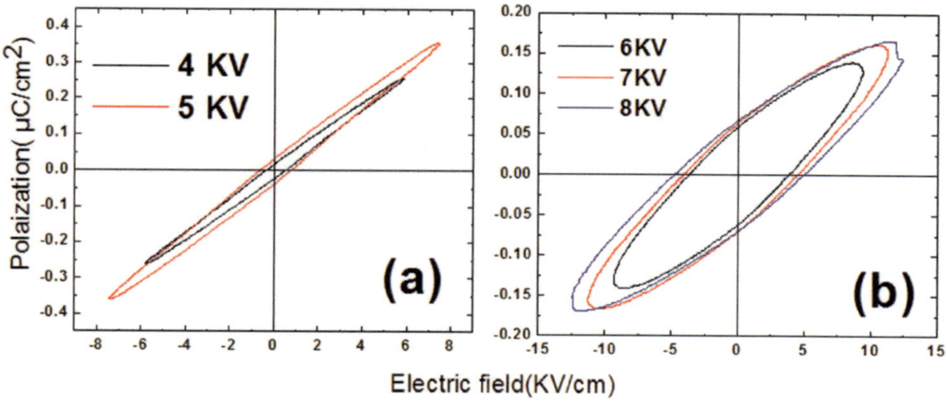

Figure 2.
(color online) polarization (P) vs. applied electric field (E) hysteresis loop measured at a frequency of 1 Hz with different voltages for pure (a) PVDF (b) n-BaTiO3.

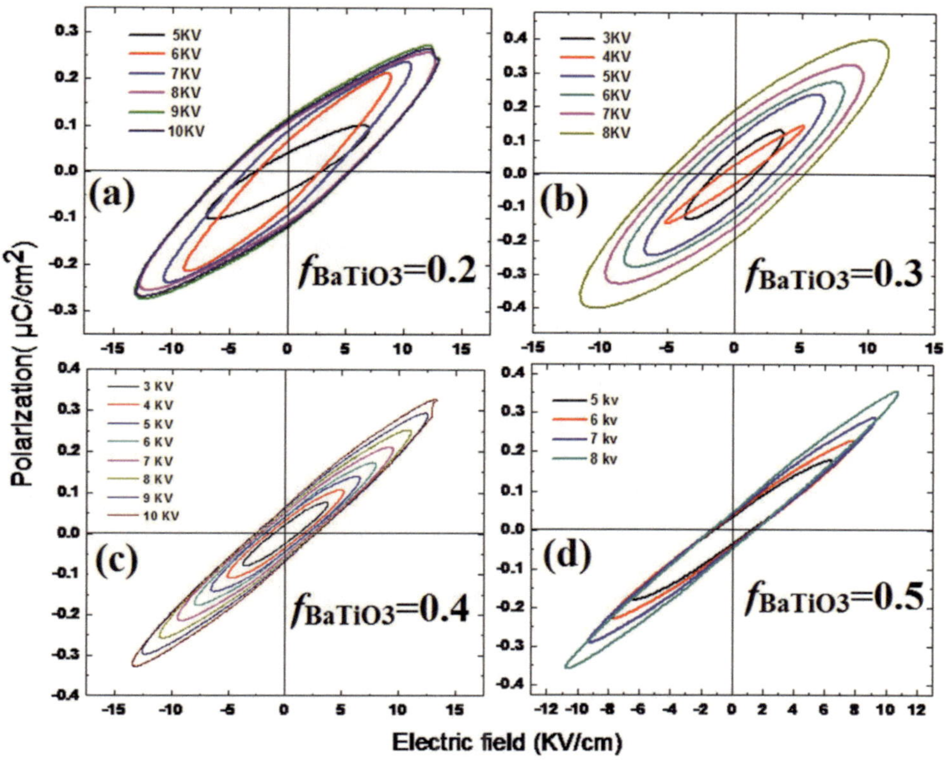

2

Figure 3.
(color online) polarization (P) vs. applied electric field (E) hysteresis loop measured at a frequency of 1 Hz with different fields for different f_{BaTiO3} *(a) 0.20 (b) 0.30 (c) 0.40 (d) 0.50.*

The experimental observation is that with increasing of the f_{BaTiO3} in the PF, the P_r, P_s and E_c also increases. This obviously indicates that the addition of *n*-BaTiO3 enhances the ferroelectric nature of the polymer material. But when the amount of *n*-BaTiO3 filler content improved, there is trivial agglomeration of filler in the PVDF matrix. The agglomeration of *n*-BaTiO3 act as hindrances, which eliminate PVDF Polymer from flowing into the BaTiO3 agglomerates and the aggregated filler causes poor enhancement in ferroelectric nature (decrement in

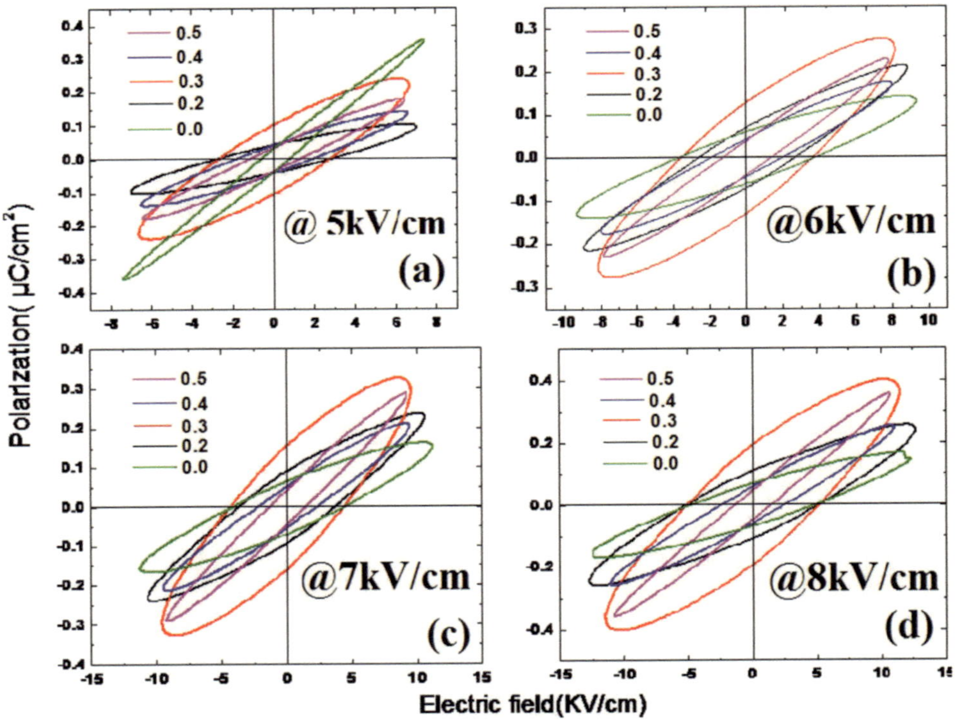

Figure 4.
(color online) polarization (P) vs. applied electric field (E) hysteresis loop measured at a frequency of 1 Hz with different f_{BaTiO3} for different fields (a) 5 kV/cm (b) 6 kV/cm (c) 7 kV/cm (d) 8 kV/cm.

the ferroelectric polarization) as evident from **Figure 4**. Conversely the dielectric properties, such as the effective dielectric constant (ε_{eff}) and loss tangent (Tan δ) of all the PF becomes a linear dependence of f_{BaTiO3} (**Figure 5**), i.e. the static ε_{eff} enhances from 10 for pure PVDF to 400 for $f_{BaTiO3} = 0.6$, whereas the loss tangent increases from 0.09 for pure PVDF to 0.9 for $f_{BaTiO3} = 0.6$ [14]. The variation in the dielectric and ferroelectric behavior is credited to the different types of structures responsible for the two altered electrical properties respectively. The dielectric properties are connected with the more interfaces in the PF, hence ε_{eff} & Tan δ enhances linearly with f_{BaTiO3} and the ferroelectric properties are associated with the ordered structure of the PF.

3.2.2 $P_s \sim f_{BaTiO3}$ for different fields

Figure 6 give you an idea about the variation of P_s as a function of f_{BaTiO3} for all the PF, at changed electric fields from 5 kV/cm to 8 kV/cm. It is practical that on increasing the concentration of *n*-BaTiO3, and also on rising the electric fields, the value of P_s raises up to $f_{BaTiO3} = 0.30$, but for $f_{BaTiO3} > 0.30$, P_s decreases, due to the aggregation of *n*-BaTiO3 causing poor improvement of the ferroelectric properties. At 5 kV/cm, the value of P_s increases from 0.1 μC/cm^2 for $f_{BaTiO3} = 0.2$ up to 0.24 μC/cm^2 for $f_{BaTiO3} = 0.30$, but when $f_{BaTiO3} > 0.30$, P_s decreases to 0.24 μC/cm^2.

3.2.3 $P_r \sim f_{BaTiO3}$ for different fields

The variation of P_r as a function of f_{BaTiO3} for all the PF (**Figure 7**), at different electric fields from 5 kV/cm to 8 kV/cm. It can be seen that on increasing the concentration of *n*-BaTiO3, P_r increases up to $f_{BaTiO3} = 0.30$, but for $f_{BaTiO3} > 0.30$,

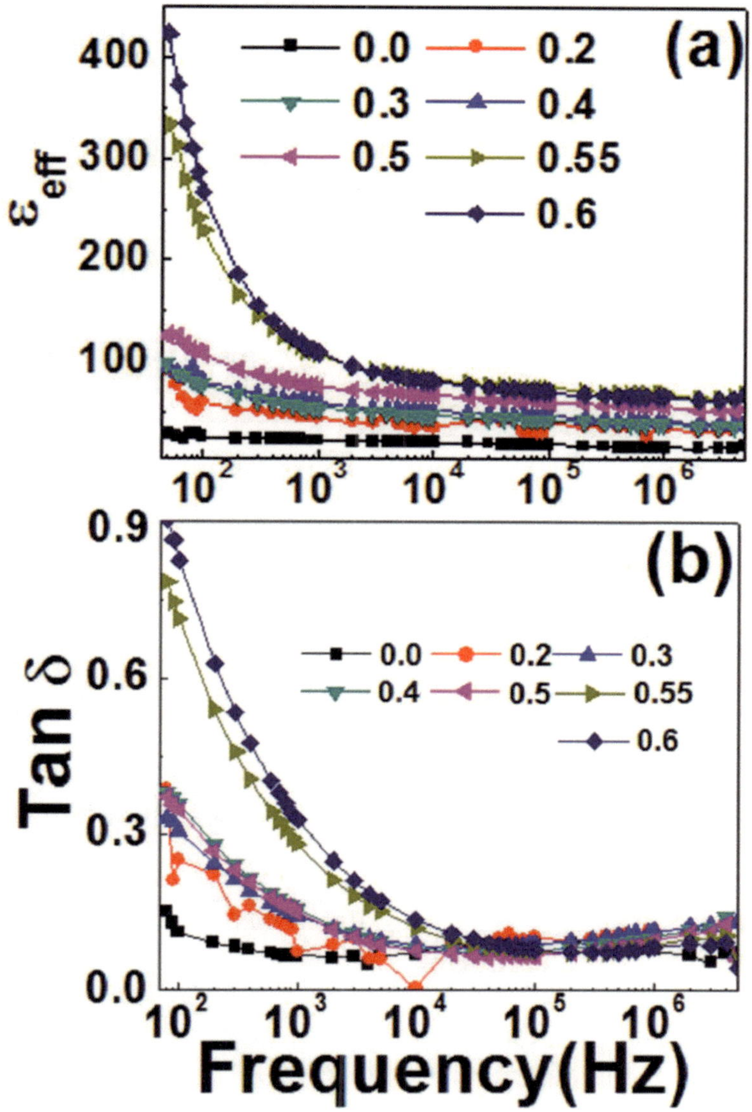

Figure 5.
(color online) the variation of (a) ε_{eff} and (b) tan δ as a function of frequency at 300 K for the PF.

P_r decreases, i.e. a parallel behavior as observed in the case of variation of $P_s \sim f_{BaTiO3}$, is also observed which is attributed to the same origin. At 8 kV/cm, P_r increases from 0.10 μC/cm^2 for f_{BaTiO3} = 0.2 up to 0.20 μC/cm^2 for f_{BaTiO3} = 0.30, but for f_{BaTiO3} > 0.30, P_r decreases and approaches to much less than 0.20 μC/cm^2.

3.2.4 $E_c \sim f_{BaTiO3}$ for different fields

The variation of E_c as a function of f_{BaTiO3} for all the PF, at different electric fields from 5 kV/cm to 8 kV/cm is shown in **Figure 8**. The value of E_c is preserved higher up to f_{BaTiO3} = 0.30, but for f_{BaTiO3} > 0.30, E_c decreases, i.e. a similar conduct as observed in the case of variation of $P_s \sim f_{BaTiO3}$, is also observed endorsed to the same origin. At 6 kV/cm, the value of E_c increases from 2.5 kV/cm for f_{BaTiO3} = 0.2 up to 3.5 kV/cm for f_{BaTiO3} = 0.30, but for f_{BaTiO3} > 0.30, E_c decreases and becomes much less than 3.5 kV/cm.

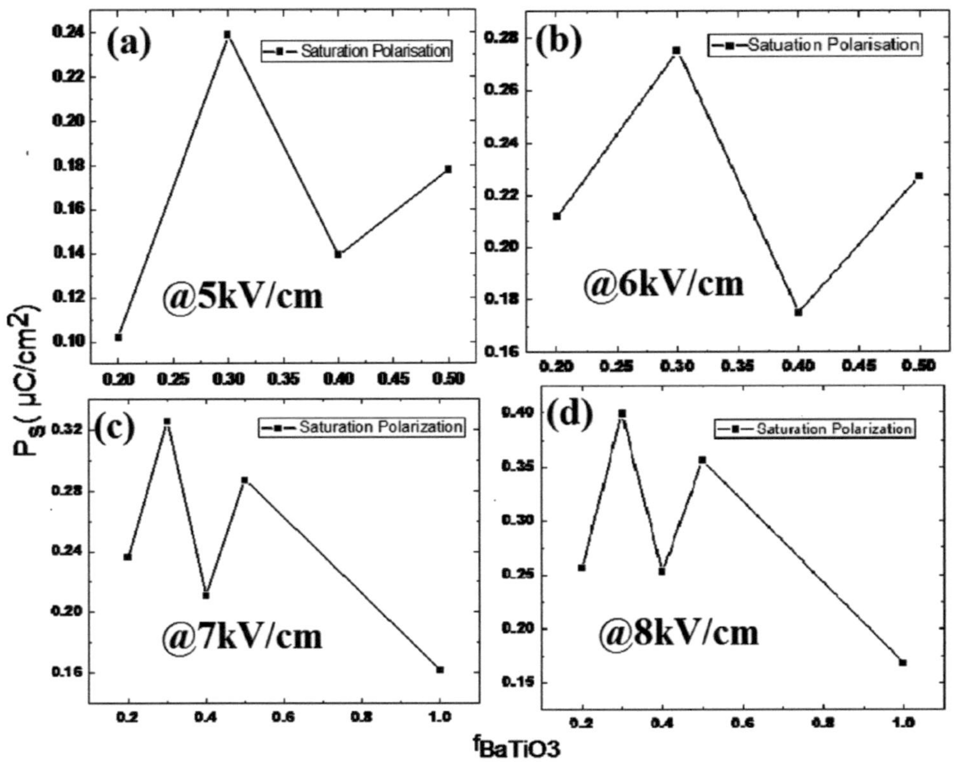

Figure 6.
$P_s \sim f_{BaTiO_3}$ at different electric fields (a) 5 kV/cm (b) 6 kV/cm (c) 7 kV/cm (d) 8 kV/cm.

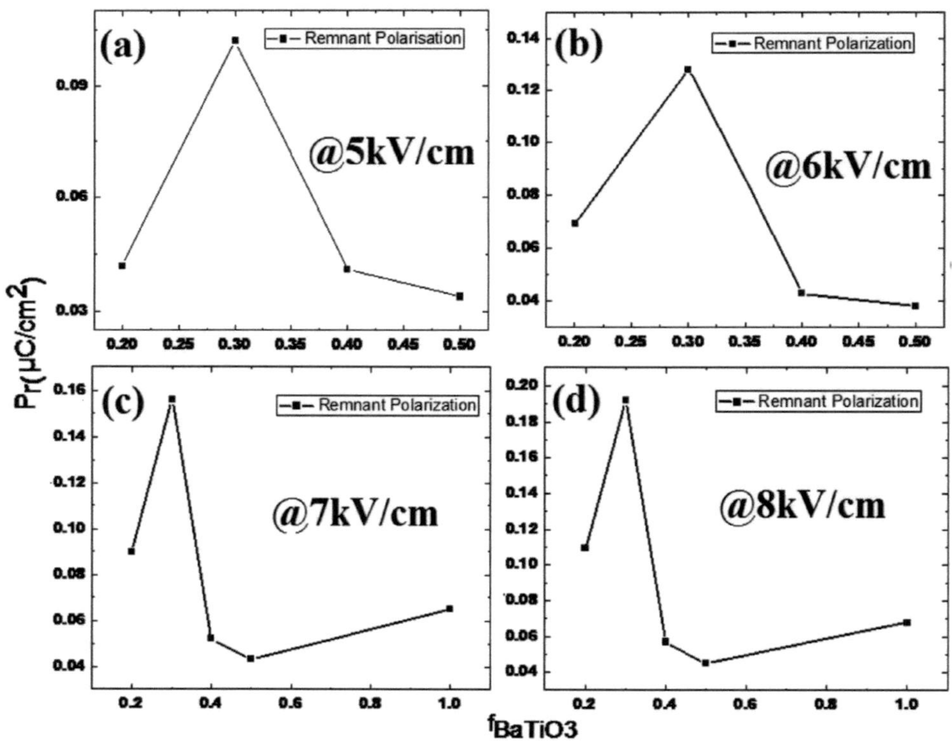

Figure 7.
$P_r \sim f_{BaTiO_3}$ at different electric fields (a) 5 kV/cm (b) 6 kV/cm (c) 7 kV/cm (d) 8 kV/cm.

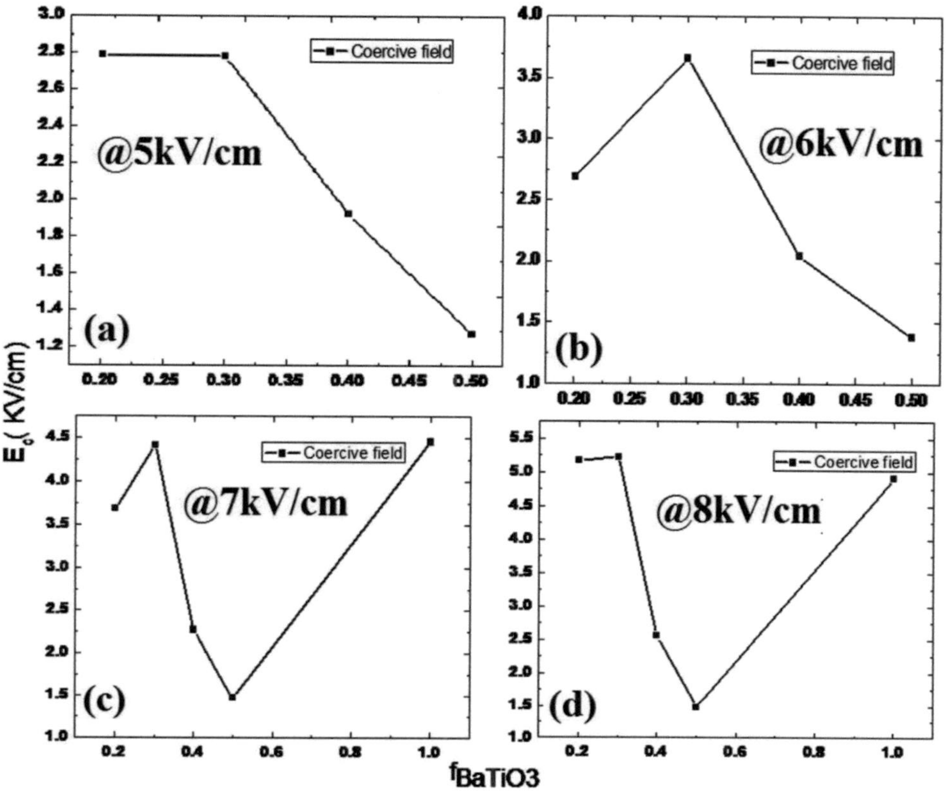

Figure 8.
$E_c \sim f_{BaTiO_3}$ at different electric fields (a) 5 kV/cm (b) 6 kV/cm (c) 7 kV/cm (d) 8 kV/cm.

3.3 Piezoelectric properties

Figure 9 gives the explanation of change in piezo- electric coefficient (d_{33}) of the PF as a function of f_{BaTiO3}. On increasing the concentration of *n*-BaTiO3, the piezoelectric nature of composite also increases and when the amount of *n*-BaTiO3 filler content increases much, the value of d_{33} becomes constant. The value of d_{33} increases from 2.10 pC/N for f_{BaTiO3} = 0.0 up to 2.20 pC/N for f_{BaTiO3} = 0.20 and remains constant, beyond f_{BaTiO3} = 0.20 up to f_{BaTiO3} = 1.0 because of the aggregated filler (*n*-BaTiO3) causes poor improvement in the peizo- electric nature of the PF.

3.4 Dielectric properties

The variation of dielectric properties of the PCC as a function of frequency at 300 K are shown in **Figure 10a** and **b**, respectively. The value of ε_{eff} at 50 Hz for the 0.0 sample is 16 while this value increases up to 120 linearly up to the PCC with f_{BaTiO3} = 0.5 & after that it raises up to the value of 330 & 420 for the samples with f_{BaTiO3} = 0.55 & f_{BaTiO3} = 0.60 respectively. The higher value of ε_{eff} for the f_{BaTiO3} = 0.55 & f_{BaTiO3} = 0.60 are attributed to the large interfacial polarization arising due to the occurrence of spherulites and created large interface like structures(during cold pressing), while the spherulites are lost for the hot molded samples (**Figure 11**).

The static dielectric constant (ε_r) of the cold pressed pure PVDF is ~16 i.e. higher than the ε_r of hot molded pure PVDF (~10) due to the loss of spherulites (Inset, **Figure 11**) of the polymer. ε_{eff} decreases with increase of frequency due to the

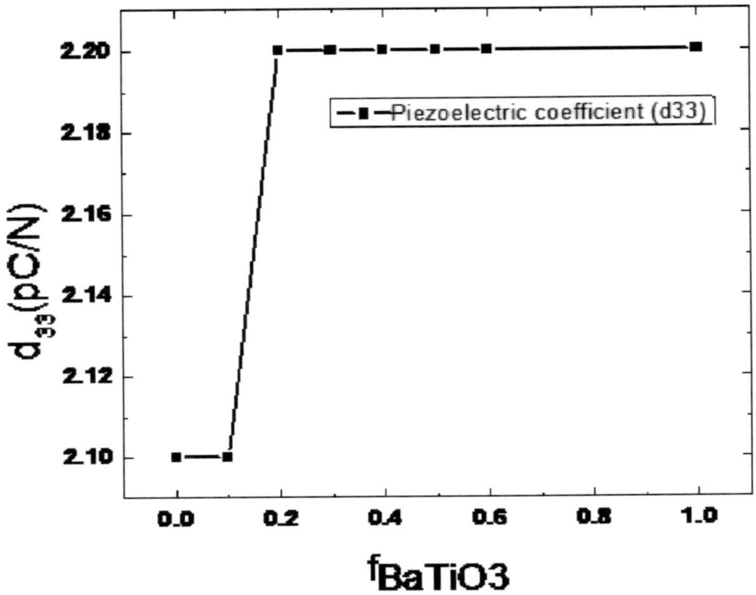

Figure 9.
Variation of the piezoelectric coefficient (d_{33}) as a function of f_{BaTiO3}.

absence of contribution from interfacial polarization at higher frequencies, where only the involvement from dipolar and atomic polarization exists. A very low Tan δ for the PCC with f_{BaTiO3} = 0.6 (having highest ε_{eff} = 420 observed at 50 Hz) was approached to 0.9 at 50 Hz and that also decreases with increase of frequency and the tendency of decrement is also observed for all the PCC. Nevertheless, in the cold pressed PMC [33], the Tan δ was reported to be 10 at 50 Hz (Inset, **Figure 10b**) for the percolative sample, with $\varepsilon_{\text{eff}} \sim 2000$. The PMC at f_c shows 10 times higher value of Tan δ in contrast to the result of PCC, although both type of polymer composites are prepared by the same cold pressing procedure. **Figure 12** shows thebehaviour of ε_{eff}, σ_{eff} and Tan δ of the composites as a function of f_{BaTiO3} at different frequencies. The ε_{eff} rises linearly from 16 to 120 for f_{BaTiO3} rises from 0.0 to 0.50 at 100 Hz, due to the large interfacial polarization occurring due to the presence of spherulites. The interfaces formed at the PCC, increases ε_{eff} largely from 120 to 350 & 420 for f_{BaTiO3} = 0.55 & 0.60 respectively. The expression developed by Yamada et al. (which is a model for explaining the sum properties of the composite) was fitted to the dielectric data (**Figure 12(b)**) at 1 kHz frequency. The model is given by

$$\varepsilon_{eff} = K_{PVDF}\left[1+\frac{nf_{BaTiO3}\left(K_{BaTiO3}-K_{PVDF}\right)}{nK_{PVDF}+\left(K_{BaTiO3}-K_{PVDF}\right)\left(1+f_{BaTiO3}\right)}\right] \tag{1}$$

where ε_{eff} is the effective dielectric constant of the composite, K_{PVDF} and K_{BaTiO3} are the dielectric constants of the polymer matrix and the ceramic, respectively, f_{BaTiO3} is the volume fraction of the ceramic and 'n' is a parameter related to the geometry of ceramic particles [2]. K_{PVDF}, K_{BaTiO3} and n found from the fitting of Eq. (1) to the dielectric data are 17, 1600 and 10, is in good agreement with the earlier literature [5].

The σ_{eff} & Tan δ increases with the increase of f_{BaTiO3} in the PCC slowly, suggesting the semiconducting nature of the BaTiO3 nano-ceramics. For f_{BaTiO3} =0.6, the σ_{eff} value varies within $10^{-8}\ \Omega^{-1}\ \text{cm}^{-1}$ to $10^{-4}\ \Omega^{-1}\ \text{cm}^{-1}$ for frequency varying between 50 Hz to 5 MHz, while the value of Tan δ is maintained in between 0.1 to 0.9.

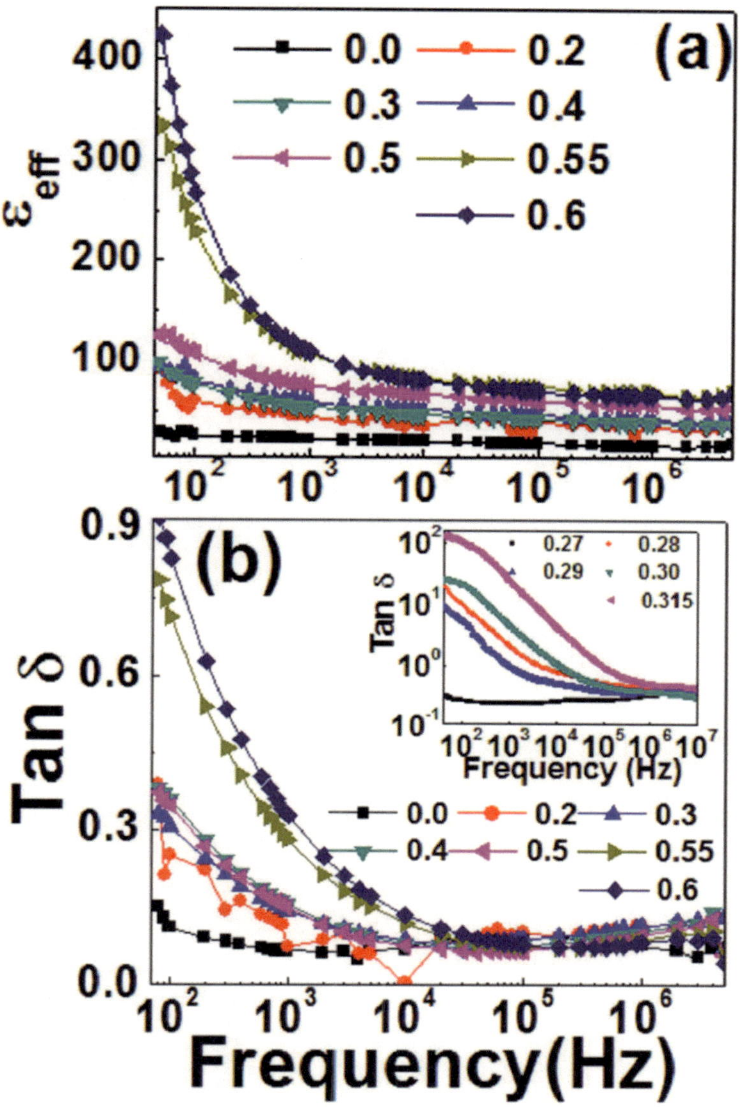

Figure 10.
(color online) the variation of (a) ε_{eff} and (b) tan δ as a function of frequency at 300 K for the PD, inset: Tan δ ~ frequency for some typical percolative PMC samples showing higher tan δ [33].

σ_{eff} & Tan δ are also found to be increasing with increase of frequency, suggesting conventional hopping conduction in the disordered PCC. Similarly, σ_{eff} value was found to be very low i.e. less than 10^{-4} Ω^{-1} for all the PCC and that value remains constant over the entire frequency range. The Tan δ raises slowly as a function of f_{BaTiO3} is found to be less than 0.9 even with the PCC having f_{BaTiO3} = 0.6.

The electrical parameters as a function of temperature of the PCC was confirmed by measuring and are given in **Figure 13**. For f_{BaTiO3} = 0.4 (**Figure 13a**) & f_{BaTiO3} = 0.50 (**Figure 13b**), the low frequency (50 Hz) value of ε_{eff} is sustained at a thermal stabilized value of 90 & 130 (with their corresponding decrement as a function of frequency) as a function of temperature from 40–100°C. The stabilization of ε_{eff} is ascribed due to the major effective contribution coming from the sum properties of the dielectric constant of both the components. Yet, for the samples with f_{BaTiO3} = 0.55 &0.6, the reached ε_{eff} ~ 350 & 420 value arises due to the sum properties

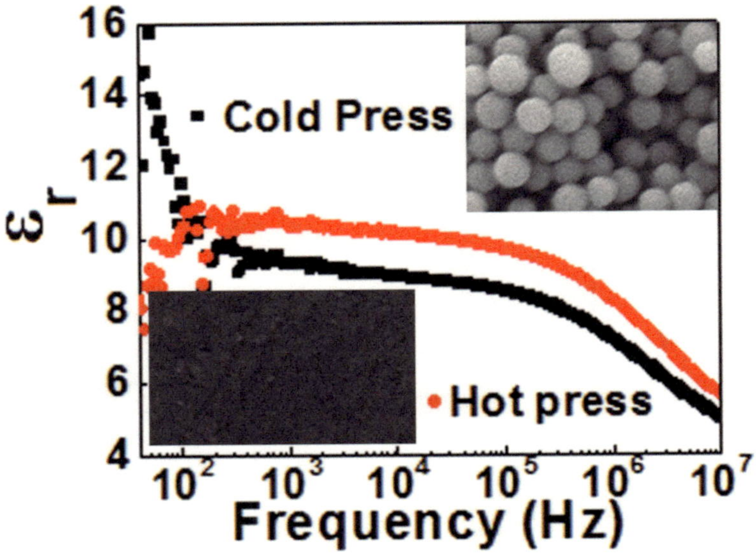

Figure 11.
The variation of dielectric constant (ε_r) with frequency for both cold and hot press PVDF, inset: The FESEM micrograph of the cold/hot molded PVDF showing the presence/loss of spherulites at temperature higher than the room temperature.

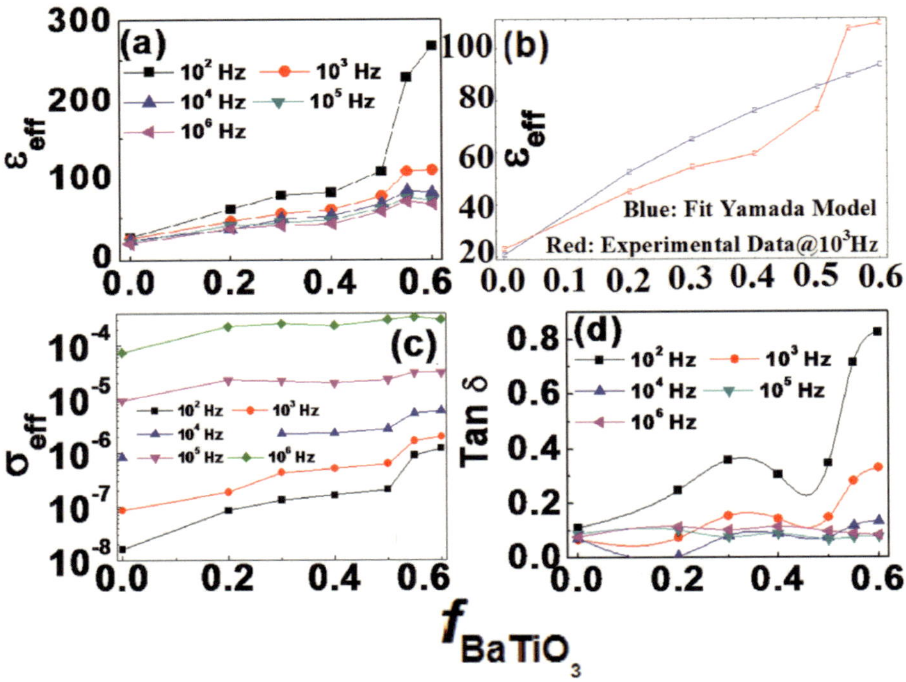

Figure 12.
(color online) the variation of (a) ε_{eff} experimentally (b) fitting of ε_{eff} experimental data at 1KHz with Yamada model as a function of f_{BaTiO_3} (c) σ_{eff} and (d) tan δ as a function of f_{BaTiO_3} for various frequencies at 300 K.

of the dielectric constant of both the components as well as also due to the major contribution of the spherulites. Hence with the rise of temperature, the ε_{eff} decreases due to the deteriorating of the spherulites of the PCC (**Figure 13c** and **d**). Hence the spherulites are useful at room temperature in case of PCC for realizing high value of ε_{eff} with lower Tan δ.

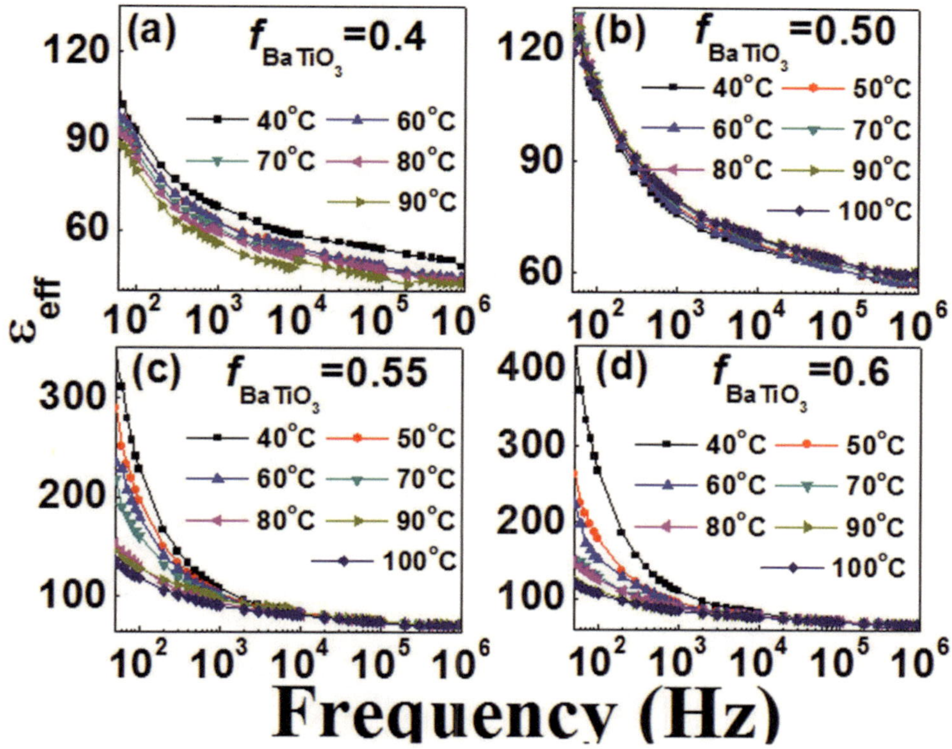

Figure 13.
(color online) the variation of ε_{eff} as a function of frequencies for the temperature varying from 40°C to 100°C for varying f_{BaTiO3} (a) 0.40 (b) 0.50 (c) 0.55 (d)) 0.60.

3.5 Electrical conductivity

Ac conductivity ($\sigma_{ac} = \omega\varepsilon_0\varepsilon Tan\delta$) as a function of frequency at different f_{BaTiO3} is shown in **Figure 14**. The σ_{eff} as a function of frequency was found to be the ac hoping conduction satisfying the Johnscher's fractional power law. The plot shows dispersion of ac conductivity with frequency corresponding to $f_{BaTiO3} \leq 0.20$, be in agreement with Eq. (2) i.e.

$$\sigma_{ac}(\omega) = \sigma_{dc} + A\omega^k \tag{2}$$

with the σ_{dc} part becoming zero and the value of $k \sim 1$. The non-presence of dc conductivity for the samples with $f_{BaTiO3} \leq 0.20$, can be understand as the non-presence of percolating paths (being formed from the semiconducting *BaTiO3* nano-ceramics in the PVDF matrix) due to insufficient fraction of *BaTiO3*. The long rage dc conduction starts to develop for f_{BaTiO3} = 0.3 to 0.5, but a good fit of Eq. (2) could not be resulted for them as the percolating paths were not sufficient. Interestingly, for $f_{BaTiO3} \geq 0.55$, a mixed conductivity is found. The plateau due to the appearance of long range dc conductivity. At higher frequency the conductivity becomes more or less with f_{BaTiO3} dependent. This "hopping or critical frequency ω_H." at which the change in slope takes place can be observed to be increasing with the increase of f_{BaTiO3}, since the length of dc plateau increases with increase of f_{BaTiO3} from 0.3 to 0.6. On the other hand, the value of '*k*' lies well within the Johnscher's universal regime [0,1] symptomatic of the validity of Johnscher's power law universally.

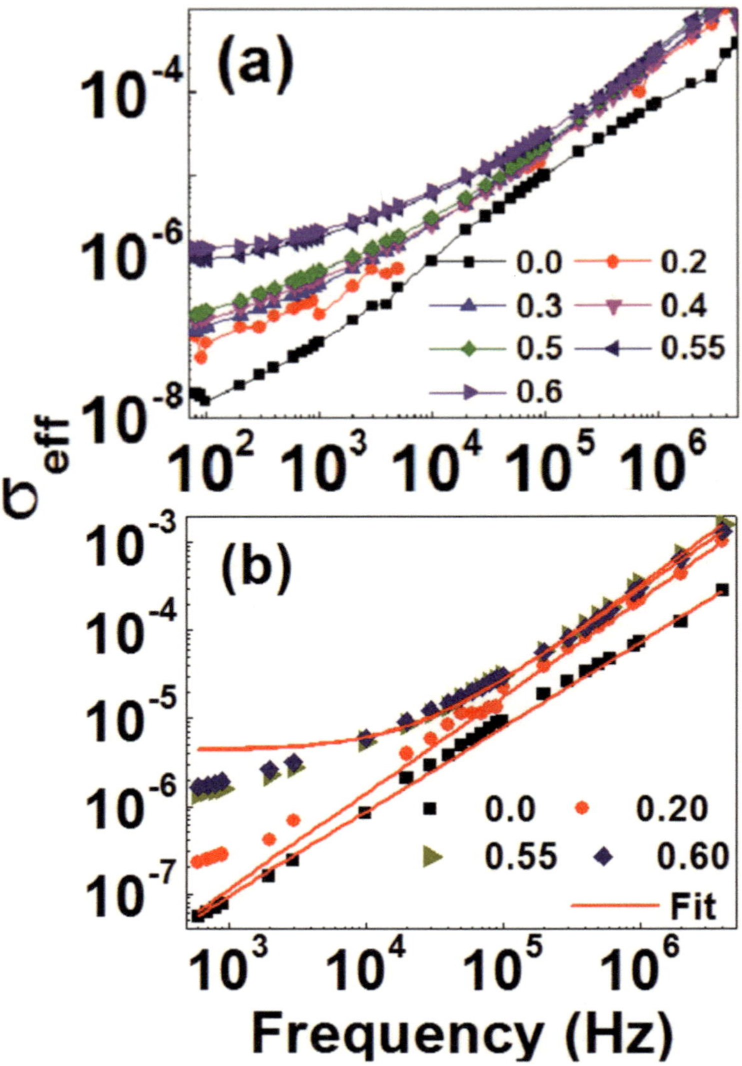

Figure 14.
(color online) the variation of (a) σ_{eff} experimentally and (b) σ_{eff} fitted with Johnscher's power law, as a function of frequencies.

3.6 Conclusions

The micro-structural, ferroelectric, piezoelectric, dielectric and conductivity properties of the polymer composites have been analyzed and are correlated. The properties strongly depend on the novel cold pressing preparation techniques and the dispersion of *n*-BaTiO3 filler particles into the PVDF matrix and also on the nano-sizes of ceramics. The addition of *n*-BaTiO3 enhances the ferroelectric, piezoelectric and dielectric properties of the composites. It is also found that this cold pressing method is more suitable to the PCC based on PVDF matrix (since very low value of Tanδ is observed). The spherulites present in PVDF matrix are always helpful in maintaining the dielectric constant and increasing the ε_{eff} of PCC. The enhancement of dielectric results are explained with the help of standard Yamada model. The mixed conductivity appears for the PD/PF and Jonschers universal fractional power law is well satisfied for all composites. The hoping conduction in these disordered materials have been confirmed in all PCC. The dynamics of charge carriers are filler/temperature dependent. These PD/PF should be explored for applications by focusing the research on achieving lowered Tan δ, which will increase the dielectric field strength/high energy density and better ferroelectric/piezoelectric properties may be expected for various multifunctional applications.

Author details

Maheswar Panda
Department of Physics, Dr. Harisingh Gour Vishwavidyalaya (A Central University), Sagar, M.P., India

*Address all correspondence to: panda.maheswar@gmail.com

References

[1] M. E. Lines, and A. M. Glass, Principles and Applications of Ferroelectrics and Related Materials, (Oxford: Clarendon Press, 2001), Vol. 680.

[2] M. Panda and A. Trivedi, Ferroelectric and Piezoelectric Properties of Cold Pressed Polyvinyledene fluoride/Barium Titanate Nano-composites, Ferroelectrics, 572, 246 (2020).

[3] G. A. Kontos, A. L. Soulintzis, P. K. Karahaliou, G. C. Psarras, S. N. Georga, C. A. Krontiras, and M. N. Pisanias, Electrical relaxation dynamics in TiO_2/ polymer matrix composites. *Expr. Poly. Lett.* **1**, 781 (2007).

[4] U.Valiyaneerilakkal, and S. Varghese, Poly (vinylidene fluoride-trifluoro-ethylene)/barium titanate nano-composite for ferroelectric nonvolatile memory devices, *AIP Advances* **3**, 042131 (2013).

[5] R. Sharma, I. P. Singh, A. K. Tripathi, and P. K. C. Pillai, Charge-field hysteresis of BaTiO3: PVDF Composites *J. Mat. Sci.* **29**, 995 (1994).

[6] M. Kuhn, and H. Kliem; Monte Carlo Simulations of Ferroelectric Properties for PVDF and BaTiO3; *Ferroelect.* **370**, 207 (2008).

[7] S. Baryshnikova, A. Milinskiya, and E. Stukovab, Dielectric properties of the ferroelectric composites [AgNa (NO2)2]0.9/[NaNO2]0.1 and [AgNa (NO2)2]0.9/[BaTiO3]0.1, *Ferroelectr.* **536**, 91 (2018).

[8] X.F.Li, H.S.Zha, P.L.Zhang, and W.L.Zhong, Composite Films OF BaTiO, and PVDF, *Ferroelectr.* **196**, 39 (1997).

[9] K. Prasad,A. Prasad,K. P. Chandra, and A. R. Kulkarni, Electrical Conduction in 0-3 BaTiO3/PVDF Composites, *Integr Ferroelectr*. **117**, 55 (2010).

[10] D. Olmos, G. Gonzalez-Gaitano, A. L. Kholkin, and J. Gonzalez-Benito, Flexible PVDF-BaTiO3 Nanocomposites as Potential Materials for Pressure Sensors; *Ferroelect.* **447,** 9 (2013).

[11] B. Hilczer, J. Kulek, M. Polomska, M. D. Glinchuk,2 A. V. Ragulya, and A. Pietraszk, Dielectric and Pyroelectric Response of BaTiO3-PVDF Nano-composites; *Ferroelect.* **316**, 31 (2005).

[12] H.-I. Hsiang, K.-Y. Lin, F.-S. Yen, and C.-Y. Hwang, Effects of particle size of BaTiO3 powder on the dielectric properties of BaTiO3/polyvinylidene fluoride composites, *J. Mater. Sci.* **36**, 3809 (2001).

[13] R. Gregorio, Jr., M. Cestari, and F. E. Bernardino, Dielectric behaviour of thin films of β-PVDF/PZT and β-PVDF/ $BaTiO_3$ composites, *J. Mater. Sci.* **31**, 2925 (1996).

[14] H. Kim, F. Torres, D. Villagran, C. Stewart, Y.G. Lin, T. Liang, and B. Tseng, 3D Printing of BaTiO3/PVDF Composites with Electric In Situ Poling for Pressure Sensor Applications, *Macromol. Mater. Eng.* **302** (11), 1700229 (2017).

[15] S. K. Mandal, S. Singh, R. Debnath, and P. Dey, Lead free xNiFe2O4 – (1-x) ErMnO3 (x = 0.1, 0.3 and 0.5) multiferroicnanocomposites: Studies of magnetoelectric coupling, AC electrical and magnetodielectric properties, *Ferroelect.* **536,** 77 (2018).

[16] S. K. Pradhan, A. Kumar, A. N. Sinha, P. Kour, R. Pandey, P. Kumar, and M. Kar, Study of ferroelectric properties on PVDF-PZT nanocomposite, *Ferroelect.* **516**, 18 (2017).

[17] M. B. Suresh, T. H. Yeh, C. C. Yu, and C. C. Chou, Dielectric and Ferroelectric Properties of Polyvinylidene Fluoride (PVDF)-Pb0.52Zr0.48TiO3 (PZT) Nano

Composite Films, *Ferroelect.* **381**, 80 (2009).

[18] H. S. Nalwa, Ferroelectric polymers, New York, Marcel Dekkar (1995).

[19] R. M. Briber, F. Khoury, The morphology of poly(vinylidene fluoride) crystallized from blends of poly (vinylidene fluoride) and poly(ethyl acrylate), *J. Polym. Sci: Polymer Physics* **31**, 1253 (1993).

[20] Q. Li, L. Chen, M. R. Gadinski, S. Zhang, G. Zhang, H. U. Li, E. Iagodkine, A. Haque, L. Q. Chen, T. N. Jackson, and Q. Wang, Flexible high-temperature dielectric materials from polymer nanocomposites, *Nature* **523**, 576-579 (2015).

[21] V. K. Thakur, and M. R. Kessler, Polymer Nanocomposites: New Advanced Dielectric Materials for Energy Storage Applications, in Advanced Energy Materials (John Wiley & Sons, Inc., Hoboken, NJ, USA, 2014).

[22] F. He, S. Lau, H. L. Chan, and J. T. Fan, High Dielectric Permittivity and Low Percolation Threshold in Nano-composites Based on Poly(vinylidene fluoride) and Exfoliated Graphite Nanoplates, *Adv. Mater.* **21**, 710 (2009).

[23] M. Panda, V. Srinivas, and A. K. Thakur, Role of polymer matrix in large enhancement of dielectric constant in polymer–metal composites, *Appl. Phys. Lett.* **99**, 0429051 (2011).

[24] M. Panda, V. Srinivas, and A. K. Thakur, Surface and interfacial effect of filler particle on electrical properties of polyvinylidene fluoride/nickel composites, *Appl. Phys. Lett.* **93**, 2429081 (2008).

[25] M. Panda, V. Srinivas, and A. K. Thakur, On the question of percolation threshold in polyvinylidene fluoride/ nanocrystalline nickel composites, *Appl. Phys. Lett.* **92**, 1329051(2008).

[26] M. Panda, V. Srinivas, and A. K. Thakur, Non-universal scaling behavior of polymer-metal composites across the percolation threshold, *Res. in Phys.* **05**, 136 (2015).

[27] Ch. Venketesh, V. Srinivas, M. Panda, and V. V. Rao, *Sol. Stat. Comm.* **150**, 893 (2010).

[28] M. Panda, V. Srinivas, and A. K. Thakur "Universal microstructure and Conductivity relaxation of Polymer-Conductor composites across the percolation threshold" *Current Applied Physics*, **14**, 1596 (2014).

[29] M. Panda, Evidence of a third kind of Johnscher's like universal dielectric response, *J. Adv. Dielect.* **8**, 1850028 (2018).

[30] M. Panda, and S. Soni, A Comparative Study of Visco-Elastic Properties of non-Polar and Polar Polymer Dielectrics" *Materials Today: Proceeding* **05**, 2123 (2018).

[31] Z.-M. Dang, H.-Y. Wang, and H.-P. Xu, Influence of silane coupling agent on morphology and dielectricproperty in BaTiO3/polyvinylidene fluoride composites, *Appl. Phys. Lett.* **89**, 112902 (2006).

[32] M. Panda, A. Mishra, and P. Shukla, Effective enhancement of dielectric properties in cold-pressed polyvinyledene fluoride/barium titanate nanocomposites, *SN Applied Sciences* **1**, 230 (2019).

[33] M. Panda, Major role of process conditions in tuning the percolation behavior of polyvinylidene fluoride-based polymer/metal composites, *Appl. Phys. Lett.* **111**, 0829011 (2017).

[34] Y. Bai, Z. Y. Cheng, V. Bharati, H. S. Xu, and Q. M. Zhang, High-dielectric-constant ceramic-powder polymer composites, *Appl. Phys. Lett.* **76**, 3804 (2000).

[35] C. K. Chiang and R. Popielarz, Polymer Composites with High Dielectric Constant, *Ferroelect.* **275**, 1 (2002).

[36] L. Zhang, D. Xiao, and J. Ma, Dielectric Properties of PVDF/Ag/ BaTiO3 Composites; *Ferroelect.* **455,** 77 (2013).

Chapter 7

Tunable Zeroth-Order Resonator Based on Ferroelectric Materials

Mohamed M. Mansour and Haruichi Kanaya

Abstract

Tunable microwave devices have the benefits of added functionality, smaller form factor, lower cost, and lightweight, and are in great demand for future communications and radar applications as they can extend the operation over a wide dynamic range. Current tunable technologies include several schemes such as ferrites, semiconductors, microelectromechanical systems (MEMS), and ferroelectric thin films. While each technology has its own pros and cons, ferroelectric thin film-based technology has proved itself as the potential candidate for tunable devices due to its simple processes, low power consumption, high power handling, small size, and fast tuning. A tunable Composite Right Left-Handed Zeroth Order Resonator (CRLH ZOR) is introduced in this chapter and it relies mainly on the latest advancement in the ferroelectric materials. It is common that for achieving optimum performance for the resonant structure, this involves the incorporation of an additional tuning by either mechanical means (i.e. with tuning screws) or other coupling mechanisms. The integration between electronic tuning and High-Temperature Superconducting (HTS) components yields a high system performance without degradation of efficiency. This leads not only low-loss microwave components that could be fine-tuned for maximum efficiency but will provide a tunable device over a broadband frequency spectrum as well. The dielectric properties of the ferroelectric thin film, and the thickness of the ferroelectric film, play a fundamental role in the frequency or phase tunability and the overall insertion loss of the circuit. The key advantages of using ferroelectric are the potential for significant size-reduction of the microwave components and systems and the cabibility for integration with microelectronic circuits due to the utilization of thin and thick ferroelectric film technology. In this chapter, ZOR is discussed and the conceptual operation is introduced. The ZOR is designed and simulated by the full-wave analysis software. The response is studied using electromagnetic characteristics with the applied electric field, ferroelectric thickness, and the operating temperature.

Keywords: HTS, Ferroelectric, Superconductor, Tunability, Zeroth-order Resonator (ZOR)

1. Introduction

Microwave devices that can be electronically tuned and switched are indispensable components for more complex and versatile communication systems. Tunable devices add new functionality and make it possible to design communication systems of reduced size and complexity. A single tunable filter, for instance, can replace a complete filter bank consisting of multiple filters. Tunable antennas arranged as an array (phase array systems) will work together as a beam antenna with beam orientation and beam angle that are electronically adjustable. Using tunable phase-shifters for separate phase-control of each antenna feed signal can accomplish this. Tunable impedance matching networks are other critical examples of devices that are required to compensate for regular and rapid changes in the radiation resistance of portable systems (e.g., cell phones).

Several technologies of providing tunability have been proposed in the literature. Tunable devices can, for example, deploy varactors and switches based on semiconductors, optical elements, liquid crystals, magnetic materials, ferroelectric materials, or microelectromechanical systems (MEMS). In terms of achievable tunability, loss contribution, linearity, tuning speed (switching rate), bias voltage, power consumption, microwave power handling capability, cross-sensitivity (temperature dependence, vibration, etc.), reliability, life cycle, area consumption, manufacturing compatibility (especially CMOS compatibility), and manufacturing cost, these approaches may differ. Therefore, the appropriate choice of the tuning method may rely on the specific application, where tunability is needed.

Each tuning mechanisms have merits and demeitrs. For instance, the magnetic and optical tuning are contactless, i.e., they do not require DC biasing network. Therefore they do not cause extra parasitics in the circuit. However, both of these methods employs a high control power. Additionally the tuning topology based on MEMS, semiconductor, etc., require DC biasing networks, which leads to wiring connection problems. The tunable metamaterials consisting of large numbers of unit cells and this causes complex wiring syste and power distribution bottleneck of the network. MEMS tuning method has additional disadvantages of being slow and requiring vacuum packaging. On the other hand, ferroelectrics have extremely low leakage currents less than MEMS counterparts. A comparison of the available tuning methods [1] shows that the ferroelectric tuning method is best suited for tunable metamaterial applications.

2. Ferroelectric tunable microwave components

Since the early 1960s, ferroelectrics have been explored for application in microwave devices and radar applications [2–5]. Their characteristics have been extensively recognized by several research groups around the world. However, their applications are starting to emerge recently [6–14]. This recent resurgent interest is due to a number of factors, such as their final application compatibility with high-temperature superconductors and similar production methods. The key to a broad variety of applications is the change in permittivity as a function of the electric field.

Frequency-agile resonators are considered a potential applications out of many devices of ferroelectrics. Such components can find a wide range of applications in several communications, industrial, commercial, and radar systems. Frequency tunability in microwave circuits can be realized using ferroelectric thin films incorporated into conventional microstrip circuits. Electronically tunable resonators can be produced with applications of interference suppression, secure communications, dynamic channel allocation, signal jamming, and ground-based communications switching. Many new systems concepts will developed as high-performance materials emerge; these systems will have considerably improved performance over conventional systems.

Ferroelectric tunable resonators have reliable perfromance, small footprint, and lightweight because they depend on electric fields and have low power consumption. The tunability factor is quite large, and devices are relatively simple in nature. The main problems currently being addressed are the relatively high loss tangents of the practical ferroelectric materials and the large bias voltages required. This may be tackled by novel device structures and superconducting conductive materials. Prior to the discussion of ferroelectric tunable resonators, it is better to discuss and demonsterate some properties of ferroelectric materials.

3. Charcterisitics of the ferroelectric materials

An attractive and efficient material that has spontaneous polarization is a ferroelectric material. The presence of spontaneous polarization is highly temperature-dependent, and ferroelectric crystals generally have phase transitions where structural changes take place in the crystal [15]. The Curie temperature (Tc) is known as this transition temperature, at which the properties of the material change abruptly.

Thermodynamic properties show large anomalies, because the nature of the crystal structure close to the Curie temperature is totally changing. This is typically the case with the dielectric constant, which increases to a high value close to the Curie temperature, as indicated in **Figure 1**; it is also the transition point where the dielectric constant has the greatest sensitivity to the application of an electric field. This important characteristic can provide an attractive deployment of such materials for tunable electromagnetic components such as resonators, antennas, power dividers, hybrid couplers, and so on.

Several materials have shown a variable permittivity with the electric field, such as Strontium titanate (SrTiO3), Barium titanate (Ba, Sr. TiO3), (Pb, Sr)TiO3, (Pb, Ca)TiO3, Ba(Ti, Sn)O3, Ba(Ti, Zr)O3 and KTaO3 dopants [16–18].

However, strontium titanate (SrTiO3, STO) and barium strontium titanate (Bax Sr.(1-x) TiO3, BSTO), where x can vary from 0 to 1, are two of the most popular ferroelectric materials currently being studied for frequency-agile components and circuits. SrTiO3 is of special interest because of its crystalline compatibility with high-temperature superconductors (HTS) and its dominant properties at low temperatures.

Pure *STO* is not supposed to have Curie temperature above 0 Kelvin. Thin films and amorphous ceramic strucutres provide a low-temperature to achieve maimum dielectric constant. This implies that the Curie temperature is above 0 Kelvin,

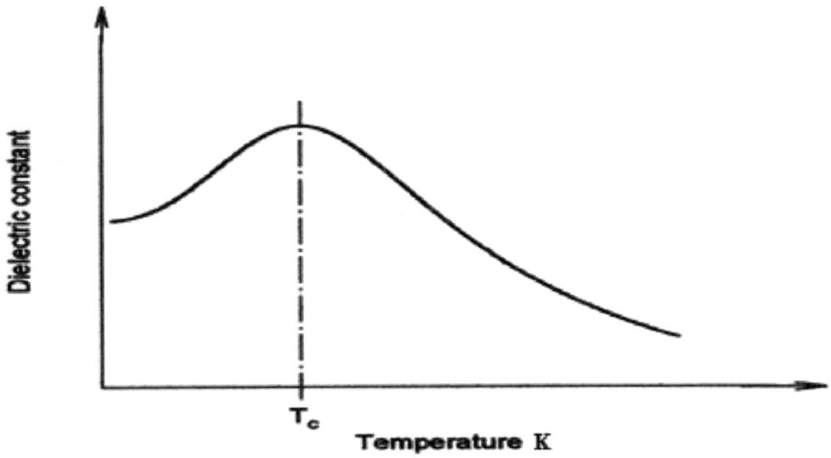

Figure 1.
Curve of dielectric constant as a function of temperature [3].

probably due to stresses or impurities in the films. For *BSTO*, as the value of x varies from 0 to 1, the Curie temperature varies from the value of *STO* to about 400 K, the Curie temperature of $BaTiO_3(BTO)$. This allows tailoring of the Curie temperature; generally, a value of $x = 0.5$ is used to optimize for room temperature, and a value of around 0.1 is used when the material is to be used in conjunction with HTS films.

For microwave and wireless communication applications, there are several different forms of ferroelectric materials that are of extrme interest. Over many years, single crystal materials have been studied. Thin-film ferroelectric materials have recently been investigated; these films have been manufactured by laser ablation and are very small in thickness, typically less than 1 μm. The films are also mainly deposited on the LaAlO-3 substrate of lanthanum aluminate (it is an inorganic substrate and was chosen to design the proposed ZOR resonator circuit) and are usually combined with single layers such as HTS or a top-surface patterned normal conductor. Tri-layer (substrate-superstrate-HTS) films, forming an HTS/Ferroelectric/HTS structure, have also been produced, however. Films on a sapphire substrate were also produced with a CeO 2buffer layer of cerium oxide to compensate for the mismatch between lattice and thermal expansion [19]. More recently, the sol–gel technique [20] has been developed for producing Barium Strontium Titanate (BST). This method can generate material that is about 0.1 mm thick.

4. Thin film Sturucres dielectric properties

The dielectric constant of bulk single-crystal STO is known to be independent of frequency up to 100–200 GHz [21–24]. The electric field and temperature dependence of the dielectric constant of single-crystal STO measured using a disk resonator at microwave frequencies [25] is shown in **Figure 2** *(The Matlab source code for calculating dielectric constant against the temperature and electric field is included in Appendix).* As can be seen, at a low temperature, the variation in the dielectric constant against an applied dc electric field is more sensitive.

5. Properties of ferroelectric materials for microwave applications

The dielectric constant for thin films is normally lower than the single crystal with the same composition, and the loss tangent can be one order higher. An example of the temperature dependence of the bulk and thin-film BST permittivity is shown in **Figure 3** of [26]. It should be noted that the permittivity is substantially lower than the bulk for the BST thin film and the sharp peak is not observed at the phase transition temperature. The effect of size or the presence of dead layers, misfit strain and thin film defects are considered to be the sources of the deviation of properties from the behavior of the bulk [27]. The theory of this deviation, however, is not well understood yet.

The most fundamental characteristics for microwave applications are the dielectric constant, tunability, and loss performance of ferroelectric materials. It is clear that for high-performance devices, high tunability and low dielectric loss are favorable. In response to the applied electric field, which is the basis of microwave applications, the dielectric constant ε of ferroelectrics varies. Tunability is a criterion for evaluating the dependence on permittivity in the electric field. The tunability of a ferroelectric material, defined as the ratio of the dielectric permittivity of the material at zero electric fields to its permittivity under electric field bias E, can be defined in two ways.

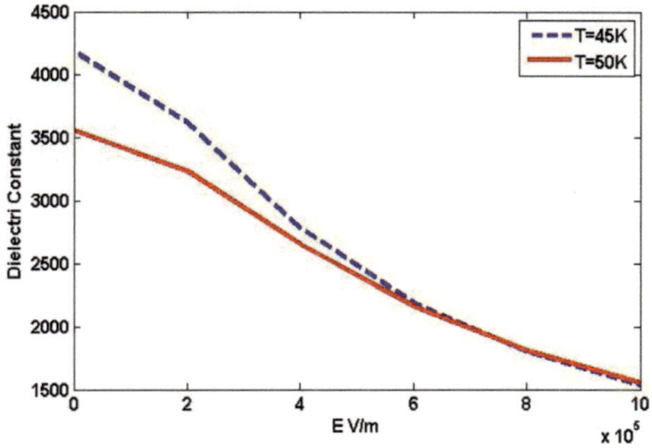

(a)

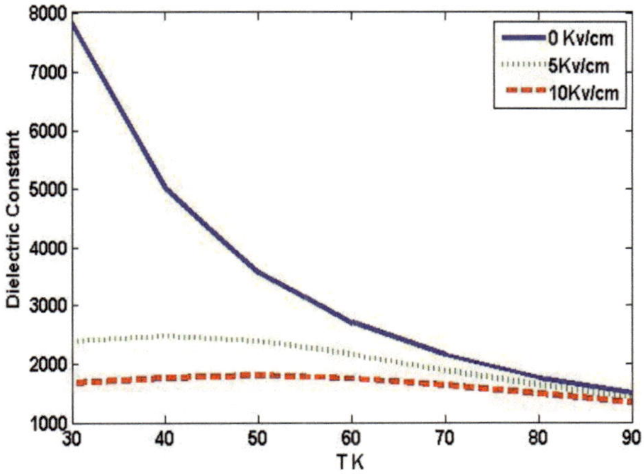

(b)

Figure 2.
(a) Electric field dependence of the STO dielectric constant (b) temperature dependence of the STO dielectric constant at different dc electric fields [25].

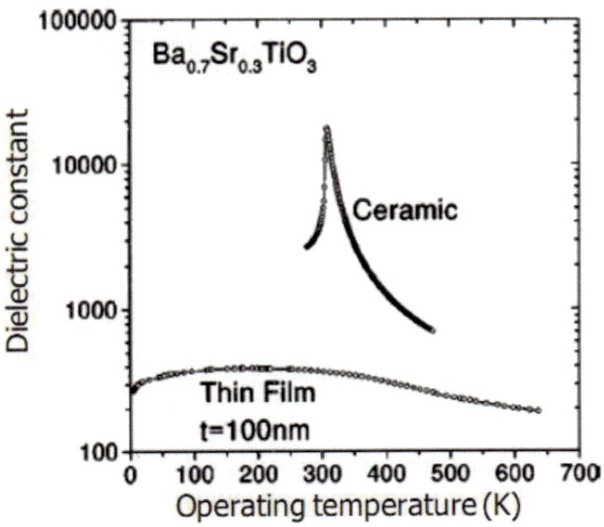

Figure 3.
Variation of the dielectric constant of a BST ceramic and thin film as a function of operating temperature [26].

$$n = \frac{\varepsilon(0)}{\varepsilon(E)} \tag{1}$$

and the relative tunability n_r defined as the relative change of the permittivity between zero bias and an electric field E with respect to its permittivity at zero bias

$$n_r = \frac{\varepsilon(0) - \varepsilon(E)}{\varepsilon(0)} = 1 - \frac{1}{n} \tag{2}$$

6. Tunable composite right/left-handed transmission lines using ferroelectric thin films

Ferroelectric/superconductor thin films ($YBa_2Cu_3O_{7-\delta}/Ba_{0.0}Sr_{0.9}TiO_3$) are used to realize an electrically tunable, low-loss composite right/left-handed transmission line. A resistive line is deployed as both the DC bias path and RF choke. The whole device maintains a simple all-planar configuration. The composite right/left-handed transmission lines that are well-matched shows a wide passband [28].

Various tuning elements, such as surface-mounted discrete varactors or ferroelectric parallel plate varactors, have been used [29]. Through the series capacitors, tuning is accomplished, although tuning the capacitance and shunt inductance [30] simultaneously would be advantageous in maintaining the impedance match for critical requirements. Cables, the introduction of RF capacitive bases, decoupled capacitors [31], or more complicated networks trace the DC bias of the reported structures.

In this study, the tunable low-loss device is realized by planar ferroelectric/superconductor thin films. Tunable elements based on ferroelectric materials avoid surface mounting components compared to discrete diode varactors, so neither contact losses nor parasitics are present in the configuration. Furthermore, it is possible to continuously vary capacitance values.

The electrical tuning is based on ferroelectric permittivity's electric-field dependency. As the fundamental tuning element for its ability, between planar structures, to establish a relatively high electric field between its two electrodes, an interdigital capacitor (IDC) is used. Between the circuit layer (HTS) and the substrate, the ferroelectric thin film is sandwiched. Using a bias tee, the DC bias can either be excited through the RF ports or through an independent bias network with an RF choke. In that it introduces less interference to the main circuit and is often easy to fit into measurement systems, the previous approach is beneficial. However, due to the high-pass nature of a CRLH-TL, it is not straightforward to apply DC voltage through the RF ports.

In this case, we propose a resistive line approach as shown in **Figure 4** for a three unit cascaded CRLH-TL. The resistive line prevent the RF trnamission from flowing into the DC path if its resistance is sufficiently high. With line dimensions of $10\mu m$ wide, the simulated responses for different surface resistances are given in **Figure 5**. The attenuation losses are estimated 0.6 dB for 20Ω surface resistance, 0.3 dB for 0Ω, and 0.2 dB for 90Ω, corresponding to a line resistance of over 2, and 9 $K\Omega$ per unit-cell.

The substrate is 0.5 mm thick MgO. The ferroelectric material ($Ba_{0.0}Sr_{0.9}TiO_3$) is deposited on the upper side of the substrate. On top of BST, the $YBa_2Cu_3O_{7-\delta}$ ($YBCO$) is deposited and represents the superconducting circuit layer. Silver is used for the contact pads to reinforce the electrical connection with the on-wafer probes and bonding wires. The resistive line is made of titanium material that has surface resistance . The BST film has a measured permittivity of 400–500 with a loss tangent of 0.02–0.03.

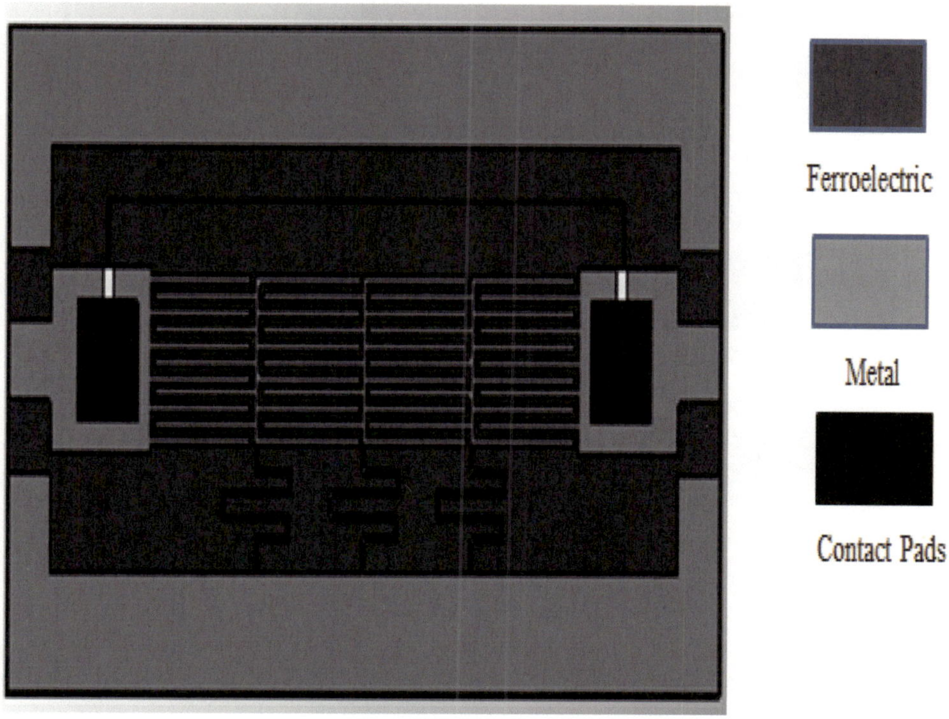

Figure 4.
Layout of a 3-unit CRLH-TL with resistive bias line, IDC, and meander line inductor [28].

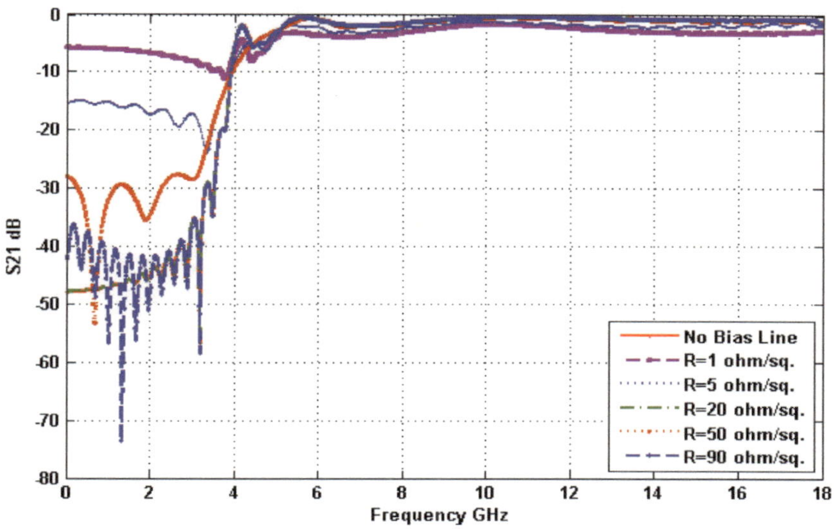

Figure 5.
Simulated responses of a 3-unit CRLH-TL with different surface resistances of the bias line.

The CRLH-TL single-unit shown in **Figure 6** was simulated and manufactured by [28]. It shows only the simulations, and the measurements are given in **Figure 7**. Within the bias range of 0-70 V, the return loss is better than −1 dB from 4 to 14GH.

Calculations show that 0.1 dB would have contributed to the ferroelectric loss ($\tan\delta = 0.03$). If the bias increases from 0 to 70 V (indicated by the effect of surface resistance variation), the phase shifts by 3.60 at 10 GHz in **Figure 8**.

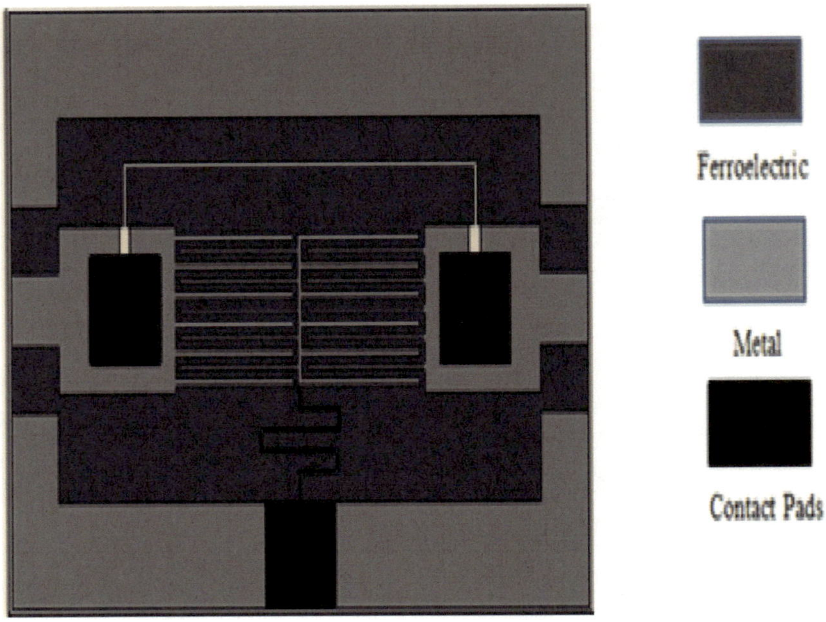

Figure 6.
Picture of the one-unit CRLH-TL [28].

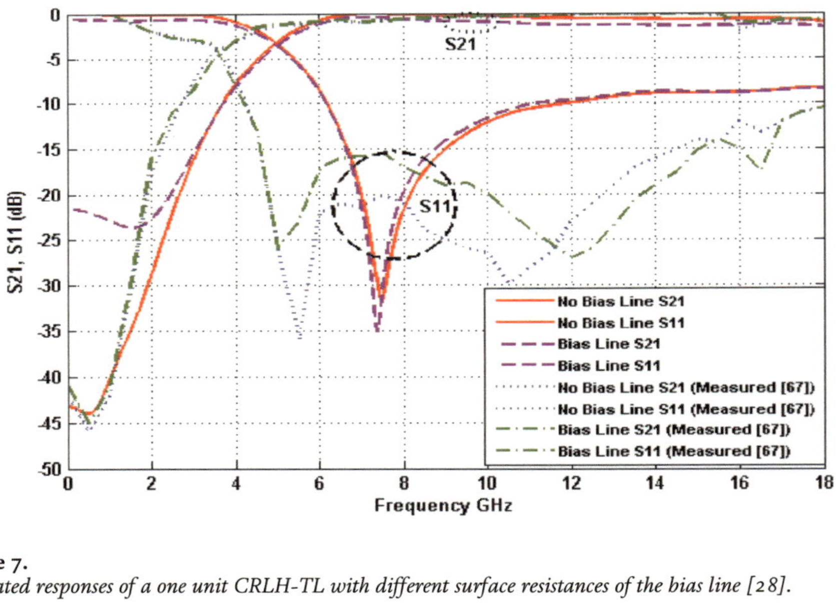

Figure 7.
Simulated responses of a one unit CRLH-TL with different surface resistances of the bias line [28].

7. Proposed tunable ZOR based on combined ferroelectric and HTS

Ferroelectric thin films can easily be incorporated into the CRLH MTM microstrip structure shown in **Figure 9**. For frequency-agile microwave communications systems, ferroelectric tunable microstrip structures are potentially attractive. These tunable components allow an innovative class of components with large frequency-agile tunabilities. Additionally, attenuation losses can be minimized when combined with HTS conductor for low-temperature applications. SrTiO3 (STO) thin films with large dielectric tunability and low loss tangents at microwave

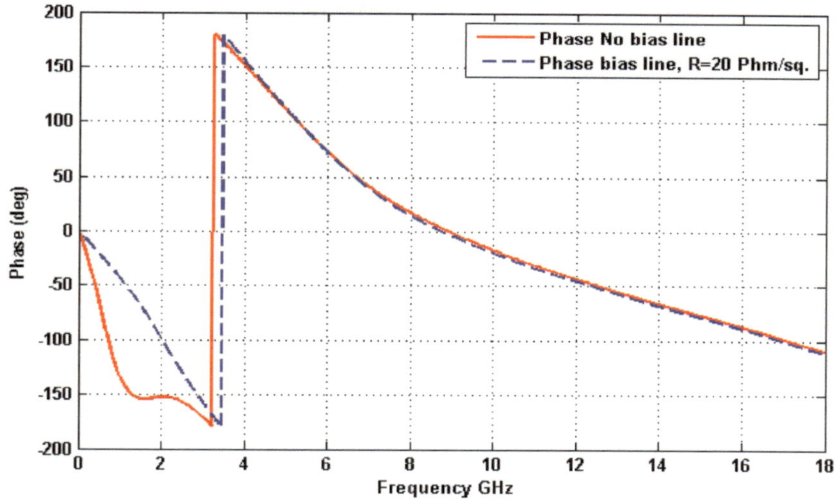

Figure 8.
Simulated phase responses and the phase shift of the one-unit CRLH-TL with two cases [28].

frequencies have been the most promising ferroelectrics for integration with HTS circuits [32]. The popular ferroelectric tunable structures are based on conductor/ ferroelectric/dielectric two-layered microstrip.

The modified microstrip structure of **Figure 9** consists of a dielectric substrate (e.g., LAO or MgO, typically 254 to 500 um thick, LAO with permittivity 23.6 is selected to represent the base substrate [32], a ferroelectric thin-film layer

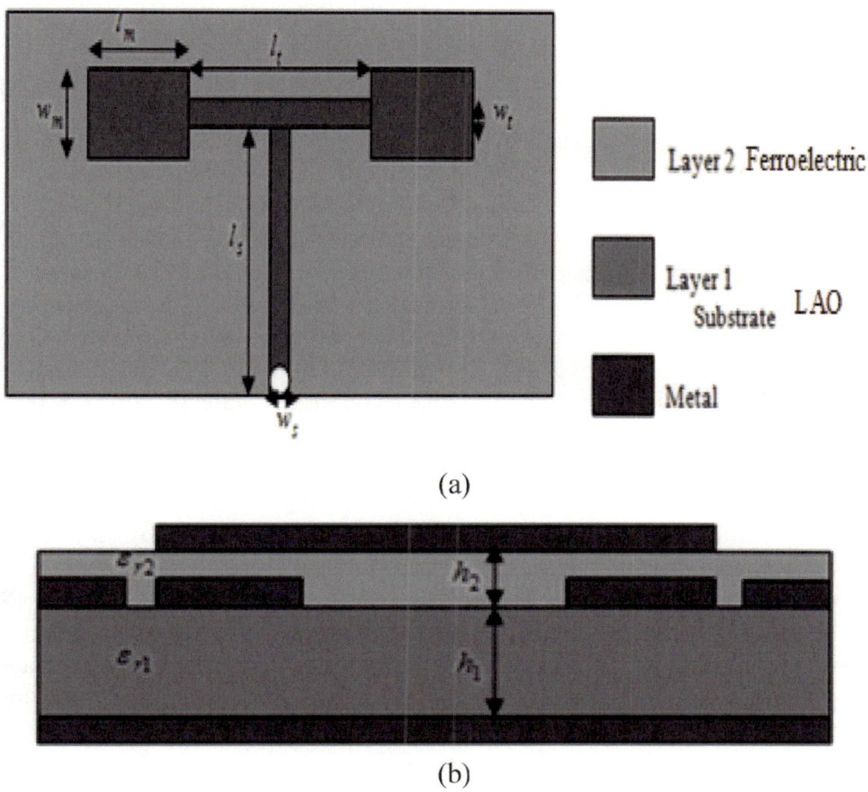

Figure 9.
Layout of ZOR with ferroelectric material (a) top view (b) side view.

(thickness."t_{ferro}" varying between 300 and 2000 *nm* for various applications), a gold or YBCO thin film (2μmthick or 300–600 *nm* thick, respectively) for the top conductor, and a 2μm thick gold ground plane. The STO film is a lossy dielectric that has a complex permittivity with a dielectric constant ε_r and a loss tangent$tan\ \delta$. Both of these parameters are functions of the DC applied electric field (E) and the temperature (T), and are introduced in the analysis by a phenomenological model developed by Vendik et al. [32] for a single crystal, and are given by [33]:

$$\varepsilon_r(T, E) = \frac{\varepsilon_{00}}{\Phi(T, E)} \tag{3}$$

$$tan\ \delta = tan\ \delta_1 + tan\ \delta_2 + tan\ \delta_3 \tag{4}$$

where

$$tan\ \delta_1 = A_1 (T/T_0)^2 / \Phi(T, E)^{\frac{3}{2}} \tag{5}$$

$$tan\ \delta_2 = A_2 \Psi(T, E)^2 / \Phi(T, E) \tag{6}$$

$$tan\ \delta_3 = \frac{A_3 n_d}{\Phi(T, E)} \tag{7}$$

and

$$\Phi(T, E) = \left[(\xi^2 + \eta^3)^{\frac{1}{2}} + \xi\right]^{\frac{2}{3}} + \left[(\xi^2 + \eta^3)^{\frac{1}{2}} - \xi\right]^{\frac{2}{3}} - \eta \tag{8}$$

$$\Psi(T, E) = \left[(\xi^2 + \eta^3)^{\frac{1}{2}} + \xi\right]^{\frac{1}{3}} - \left[(\xi^2 + \eta^3)^{\frac{1}{2}} - \xi\right]^{\frac{1}{3}} \tag{9}$$

$$\xi(E) = \sqrt{\xi_s^2 + (E/E_N)^2} \tag{10}$$

$$\eta(T) = \left(\frac{\Theta}{T_0}\right) \sqrt{(1/16) + (T/\Theta)^2} - 1 \tag{11}$$

In the previous equations, ε_{00} is a constant analogous to the Curie constant, E_N is the normalizing applied electric field, ξ_s is the rate of crystal strain, a measure of the density of defects, Θ is the effective Debye temperature, and T_0 is the effective Curie temperature. Numerical values for these model parameters for a single crystal of STO are given by [34]: ε_{00}= 2080, E_N = 19.3 KV/cm, ξ_s= 0.018, Θ= 17 K, and T_0= 42 K. The change of dielectric constant with frequency is generally small in the microwave frequency range. In Eqs. (5), (6) and (7) A_1, A_2, *and* A_3 are material parameters, and n_d is the density of charged defects. For high-quality crystals, tanδ_3 is small with respect to tan δ_1 and tan δ_2 and can be neglected. Numerical values for A_1 *and* A_2 parameters for STO at a frequency of 10 GHz are given by [35]: $A_1 = 2.4 * 10^{-4}$ and $A_2 = 4 * 10^{-3}$.

Figure 10 depicts the magnitude of the transmission scattering parameter (S21) for 0ΩYBCO/STO/LAO microstrip line CRLH ZOR shown in **Figure 9**, with a 0.3 μm STO thin film. It provides the STO permittivity dependence on its thickness. **Table 1** summarizes the plotting curves given in **Figure 10(a, b)**. As we see, for a given frequency, the attenuation increases with film thickness. At higher frequencies, because of the skin depth effect, more RF field is concentrated in the ferroelectric film and less is concentrated in the dielectric substrate, resulting in larger insertion loss. As the value of the ε_r increases, the attenuation also increases. This is a consequence of mismatches resulting from the decrease in Z_o.

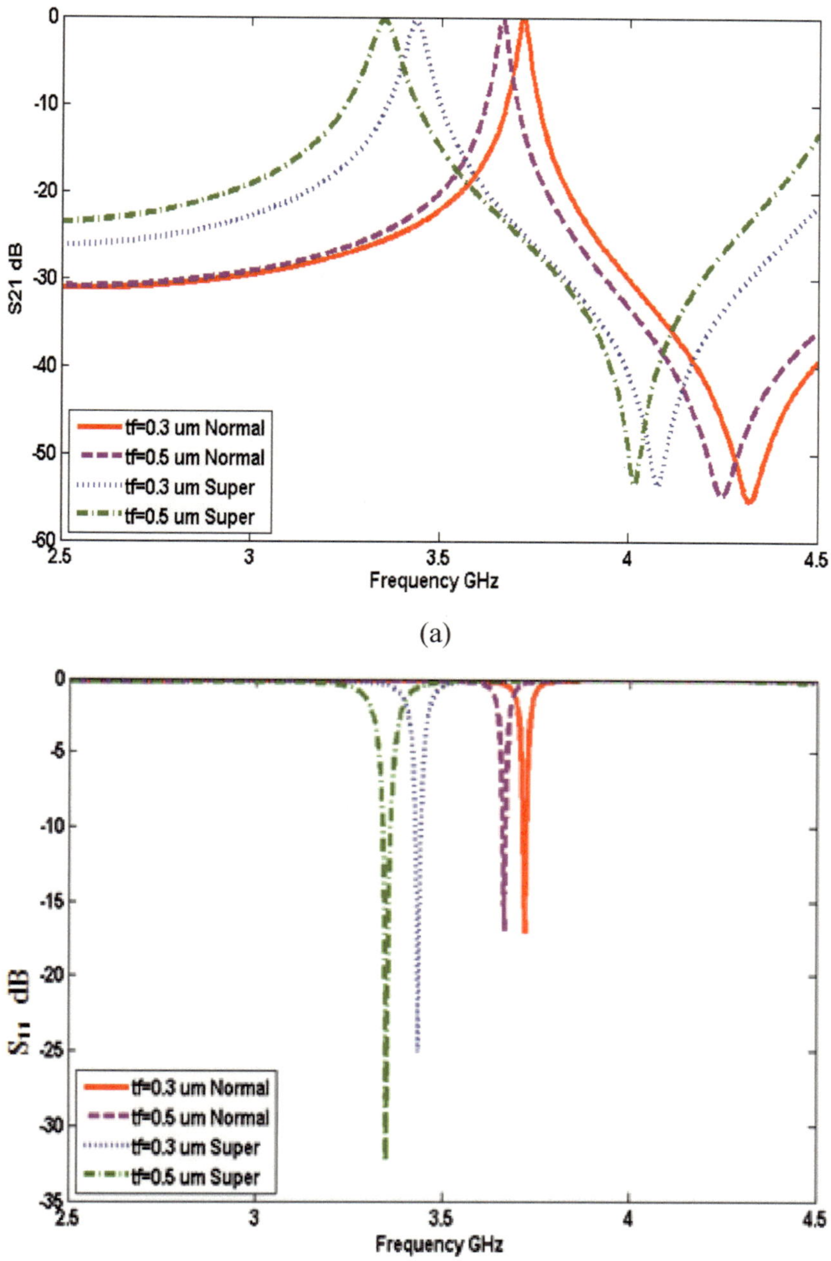

(b)

Figure 10.
Variation of the magnitude of the transmission coefficient with the frequency at a different ferroelectric thickness t_{ferro}. *(a)* S_{21}. *(b)* S_{11}.

t_{ferro} (μm)	Normal 300 K		Super 77 K	
	$f_{res}(GHz)$	S_{21} dB	$f_{res}(GHz)$	S_{21} dB
0.3	3.71	−0.95	3.43	−0.29
0.5	3.66	−0.97	3.5	−0.34

Table 1.
Parameters values of ***Figure 10***.

Temperature (°c)	ε_r	$tan\delta$ *10^{-4}
−50/223 K	438	4.1
−25 /248 K	380	4.5
0/273 K	3 0	4.8
+25/298 K	320	5.2

Table 2.
Dielectric constant and loss tangent of single-crystal SrTiO3 material at various temperatures.

$T(K)$	ε_r	$f(GHz)$	S_{21} dB
30	7811	3.41	−0.25
60	269	3.5 3	−0.21
77	1867	3.56	−0.22

Table 3.
*Parameters values of **Figure 11**.*

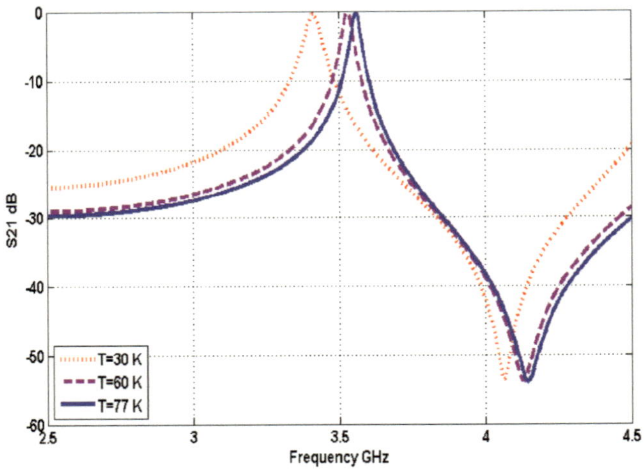

(a)

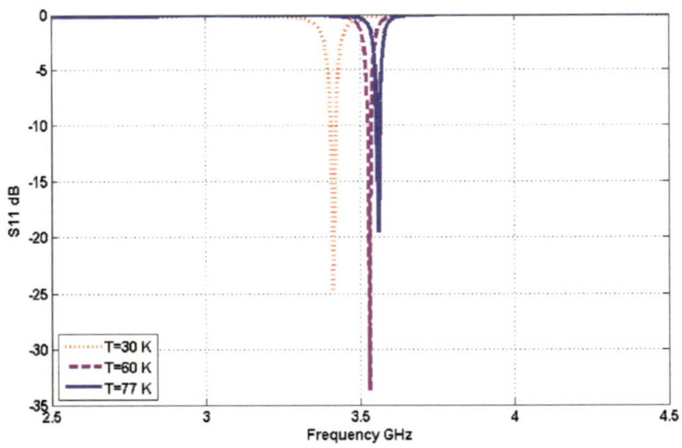

(b)

Figure 11.
Variation of the magnitude of the transmission coefficient with the frequency at different operating temperature T. (a) S_{21}. (b) S_{11}.

The temperature dependence of ε_r and $tan\ \delta$ for single-crystal SrTiO3 ferroelectric material has been reported by Krupka et al. [36] and is summarized in **Tables 2** and **3**.

Figure 11 shows the variation of resonance frequency with the temperature which is very attractive at very low temperatures.

The data were conducted at 77 K in the 2.5 to 4.5 GHz frequency range. It is worth to observe that, at no bias and at 3 GHz, the insertion loss IL = 0.2 dB. These simulated data are not de-embedded, meaning that any contribution from the SMA launchers used for the measurement to the overall insertion loss has not been compensated from the data.

Figure 12 and **Table 4** indicates the effect of E-field on the response of ZOR, the E-fields is varying from 0 to 30 Kv/cm. If the sample kept at a constant temperature

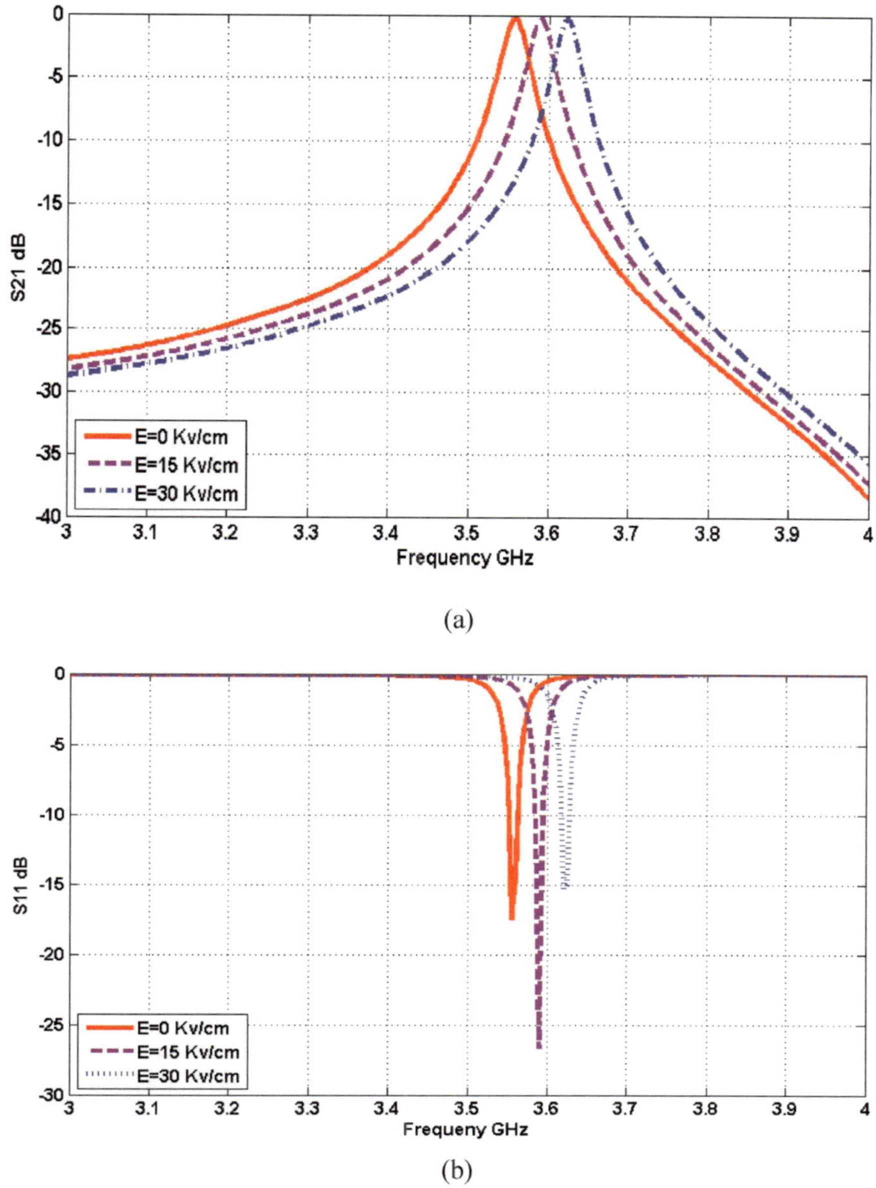

Figure 12.
Variation of the magnitude of the transmission coefficient with the frequency at different electric field E. (a) S_{21} (b) S_{11}.

$E(Kv/cm)$	ε_r	$\tan\delta$ *10^{-4}	$f(GHz)$	S_{21} dB
0	1867	7.2	3.55	−0.25
1	1168	18.5	3. 59	−0.28
30	741	26	3.62	−0.39

Table 4.
*Parameters values of **Figure 12** at T = 77 K.*

Property		MIM	HTS 77 k [Proposed]	Ferro+ Copper [Proposed]	Ferro+ HTS 77 k [Proposed]
Performance	S_{11}	−10	−34	−18	−32
	S_{21}	−3.55	−0.09	−0.54	−0.023
Tunability		No	No	Yes	Yes
Q factor		1762	43411	9957	68714
Fabrication		Simple	Difficult	Simple	Difficult

Table 5.
A comparative study of the different ZOR structures.

of 77 K, the ε_r STO is reduced from a high value of approximately 1867 at zero bias to a lower value of 741 at a high bias field.

The dynamic range of dielectric tunability with low additional microwave dielectric losses due to the insertion of ferroelectric thin films is one of the important criteria for the use of ferroelectric thin films in tunable circuits. Dielectric tunability is defined as the ($\varepsilon_r(0)$ at zero bias – $\varepsilon_r(E)$ at large bias)/ ε_r at zero bias.

$$n = \frac{\varepsilon_r(0) - \varepsilon_r(E)}{\varepsilon_r(0)} \tag{12}$$

Dielectric tunability as high as 90% is attainable in STO thin films at moderate loss-tangent values (typical values between 0.005–0.01 at GHz frequencies) [37]. So, this structure will provide a tunability up to 47% for E = 30 kV/cm.

Table 5, demonstrates the comparison between the four proposed structures stated as:

Case 1: The MIM with a normal conductor (Copper).
Case 2: Adding HTS in replace of Copper to the MIM.
Case 3: Adding the ferroelectric material in case of Copper.
Case 4: Adding the ferroelectric material in case of HTS.

8. Conclusions

A tunable ZOR CRLH resonator is successfully illustrated using a thin film ferroelectric material. ZOR can be applied to the ferroelectric material and this provides a different resonance frequency by altering either the electric field applied or the operating temperature. In addition, the incorporation of HTS material in place of normal conductors (e.g., gold, copper) significantly reduced conductor losses and consequently improved circuit performance. Therefore, for the development of low-loss and tunable microwave components and systems for wireless, radar and satellite communications, the design, manufacture and optimization of

HTS/ferroelectric hybrid circuits may be of great interest. Tunable HTS/STO/LAO ZOR resonators have a more than 47 percent frequency tunability factor and have been demonstrated at 77 K.

Acknowledgements

The author is deeply grateful to colleagues and professors at the Faculty of Electronic Engineering, Monofiya University, Egypt, for contributing and insightful comments and suggestions in conducting the above research. I would also like to express my gratitude to Kyushu University, Japan, and Prof. Haruichi Kanaya for his constructive revision, supervision, and financial support. Without their assistance and consultation, the research work would not come true. This work is partially supported by the Cabinet Office (CAO), Crossministerial Strategic Innovation Promotion Program (SIP), "An intelligent knowledge processing infrastructure, integrating physical and virtual domains" (NEDO), SCOPE, and KAKENHI (JP18K04146) JSPS.

Conflict of interest

No conflict of interest.

Appendices

Ferroelectric Characterization:

```
clear all
close all
clc
ck=8.7e+4; % Curie constant
e0=8.85e-12;
dn=2;
es=0.018;theta=175;a1=2.45e-4;t0=42;a2=4e-3; % Constants
e00=ck/t0;
En=(2*dn)/((e0)*(3*e00)^(3/2));
t=[77] %operating temperature
for j=1:1:length(t)
eta=((theta/t0)*sqrt((1/16)+(((t(j))/(theta))^2))-1) %eta(T)
  E=0:1e5:30e5 %operating electric field
for i=1:1:length(E)
    e(i)=sqrt((es^2)+((E(i)/En)^2)); %e(E)
    x(i)=(((e(i)^2)+(eta^3))^0.5)+(e(i));
    y(i)=(((e(i)^2)+(eta^3))^0.5)-(e(i));
    phi(i)=((x(i))^(2/3))+((y(i))^(2/3))-(eta);%phi(T,E)
    er(i,j)=(e00)/(phi(i)) %permittivity
    epsi(i)=((x(i))^(1/3))-((y(i))^(1/3)); %epsi(T,E)
    delta2(i)=((a2)*(epsi(i))^2)/(phi(i)); % tan delta2
    delta1(i,j)=((((a1)*(t(j)/t0)^2))/(phi(i))^(3/2)); %tan delta1
    delta(i,j)=delta1(i,j)+delta2(i)          % Total Tangetial loss
    n=er/er(i)

end
```

```
      plot(E,er)

   end
    figure
      plot(E,delta)
```

Author details

Mohamed M. Mansour[1,2]* and Haruichi Kanaya[1]

1 Electrical Engineering, Kyushu University, Fukuoka, Japan

2 Electronics Research Institute, Giza, Egypt

*Address all correspondence to: m.mansour01@ieee.org; m.mansor85@eri.sci.eg

References

[1] Z.-Y. Shen, C. Wilker, P. Pang, D. W. Face, C. F. Carter, and C. M. Harrington, "Power Handling Capability Improvement on High-Temperature Superconducting Microwave Circuits," IEEE Trans., Applied Superconductivity, 7, June 1997, 2446–24 3.

[2] O. G. Vendik, L. T. Ter-Martirosyan, A. I. Dedyk, S. F. Karmanenko, and R. A. Chakalov, "High T_c superconductivity: New applications of Ferroelectrics at Microwave Frequencies," Ferroelectrics, 144, pp. 33-43, 1993

[3] A. M. Hermann, R. M. Yandrofski, J. F. Scott, A. Naziripour, D. Galt, J. C. Price, J. Cuchario, and R. K. Ahrenkiel, "Oxide superconductors and ferroelectrics materials for a new generation of tunable microwave devices," J. Superconduct., 7, no. 2, pp. 463-469, 1994

[4] A. T. Findikoglu, Q. X. Jia, X. D. Wu, G. J. Chan, T. Venkatesan, and D. W. Reagour "Tunable and Adaptive Bandpass Filter Using Non-linear Dielectric Thin Film of SrTiO3" Appl. Phys. Lett., vol. 64, pp. 16 1–16 3, 1996.

[5] A. Kozyrev, A. Ivanov, V. Keis, M. Khazov, V. Osadchy, T. Samoilova, O. Soldatenkov, A. Pavlov, G. Koepf, C. Mueller, D. Galt, T. Rivkin, "Ferroelectric films: Nonlinear properties and applications in microwave devices," IEEE MTT-S, Digest, pp. 98 –988, 1998.

[6] G. Subramanyam, F. W. Van Keuls, and F. A. Miranda "A novel K-band Tunable Microstrip Bandpass Filter Using A thin Film HTS/Ferroelectric/Dielectric Multilayer Configuration," IEEE MTT-S, Digest, pp. 1011–1014, 1998.

[7] G. Subramanyam, F. W. Van Keuls, and F. A. Miranda, "A K-band tunable microstrip bandpass filter using a thin film conductor/ferroelectric/dielectric multilayer configuration," IEEE Microwave and Guided-Wave Lett., vol. 8, pp. 78–80, June 1998.

[8] F. A. Miranda, G. Subramanyam, F. W. Van Keuls, R. R. Romanofsky, J. D. Warner, and C. H. Mueller, "Design and development of ferroelectric tunable microwave components for Ku- and K-band satellite communication systems," IEEE Trans., MTT-48, pp. 1181–1189, July 2000.

[9] M. J. Lancaster, J. Powell, and A. Porch, "Thin-film ferroelectric microwave devices," Supercond. Sci. Technol., vol. 11, pp. 1323–1334, 1998.

[10] R. Romanofsky, J. Bernhard, G. Washington, F. VanKeuls, F. Miranda, and C. Cannedy, "A K-band Linear Phased Array Antenna Based on $Ba_{0.6}Sr_{0.4}TiO_3$ Thin Film Phase Shifters," IEEE MTT-S, Digest, pp. 13 1–13 4, 2000.

[11] A. Kozyrev, V. Osadchy, A. Pavlov, L. Sengupta, "Application of Ferroelectrics in Phase Shifter Design," IEEE MTT-S, Digest, pp. 13 –13 8, 2000.

[12] G. Subramanyam, F. A. Miranda, R. Romanofsky, F. W. Van Keuls, C. L. Canedy, S. Aggarwal, T. Venkatesan, and R. Ramesh, "A ferroelectric tunable microstrip Lange coupler for K-band application," IEEE MTT-S, Digest, pp. 1363–1366, 2000.

[13] I. Vendik, O. Vendik, V. Sherman, A. Svishchev, V. Pleskachev, A. Kurbanov, and R. Wordenweber, "Performance limitation of a tunable resonator with a ferroelectric capacitor," IEEE MTT-S, Digest, pp. 1371–1374, 2000.

[14] Y. Xu, Ferroelectric Materials, Elsevier, New York, 1991.

[15] S. B. Herner, F. A. Selmi, V. V. Varadan, and V. K. Varadan, "The effect of various dopants on the dielectric properties of barium strontium Titinate," Mater. Lett. 1, pp. 317–324, 1993.

[16] C. M. Jackson, J. H. Kobayashi, A. Lee, C. Prentice-Hall, J. F. Burch, and R. Hu, "Novel monolithic phase shifter combining ferroelectrics and high temperature superconductors," Microwave Opt. Technol. Lett., pp. 722–726, 1992.

[17] R. C. Neville, B. Hoeneisen, and C. A. Mead, "Permittivity of Strontium Titinate," J. Appl. Phys. 43, pp. 2124–2131, 1972.

[18] Y. A. Boikov and T. Claeson, "High Tunability of The Permittivity of $YBa_2Cu_3O_7/SrTiO_3$ Hetro-Structures on Sapphire Substrates," J. Appl. Phys. 81, no. 7, pp. 3232–3236, 1997.

[19] F. De Flavis, D. Chang, N. G. Alexopoulos, and O. M. Stafsudd, "High purity ferroelectric materials by sol–gel process for microwave applications," IEEE MTT-S, Digest, pp. 99–102, 1996.

[20] O. G. Vendik, "Dielectric nonlinearity of Displacive ferroelectrics at UHF," Ferroelectrics, vol. 12, pp. 8 –90, 1976.

[21] R. C. Neville, B. Hoeneisen, and C. A. Mead, "Permittivity of Strontium Titinate," J. Appl. Phys. 43, pp. 2124–2131, 1972.

[22] S. Gevorgian, E. Carlsson, P. Linner, E. Kollberg, O. Vendik, and E. Wikborg, "Lower order modes of YBCO/STO/ YBCO circular disk resonators," IEEE Trans., MTT-44, pp. 1738–1741, Oct. 1996.

[23] O. G. Vendik, E. K. Hollmann, A. B. Kozyrev, and A. M. Prudan, "Ferroelectric tuning of planar and bulk microwave devices", Journal of Superconductivity, vol. 12, pp. 32 -338, 1999.

[24] A. K. Tagantsev, V. O. Sherman, K. F. Astafiev, J. Venkatesh, and N. Setter, "Ferroelectric Materials for Microwave Tunable Applications," Journal of Electroceramics, vol. 11, pp. -66, 2003.

[25] N. K. Pervez, P. J. Hansen, and R. A. York, "High Tunability Barium Strontium Titanate Thin Films for RF Circuit Applications," Applied Physics Letters, vol. 8, no. 19, pp. 44 1-44 3, 2004.

[26] R. N. Simons, Coplanar waveguide circuits, Components, and Systems, John Wiley & Sons, USA, chapter 12, 2001.

[27] D. Kuylenstierna, A. Vorobiev, P. Linner, S. Gevorgian", Composite right/ left handed transmission line phase shifter using ferroelectric Varactors," IEEE Microw. Wireless Compon. Lett., vol. 16, no. 4, pp. 167-169, 2006.

[28] Yi Wang, Michael J. Lancaster, Frederick Huang, Phe. M. Suherman, Donna M. Holdom, Timothy J. Jackson, "Superconducting Tunable Composite Right/Left-Handed Transmission Lines Using Ferroelectric Thin Films with a Resistive Bias Network", IEEE MTT-S International Microwave Symposium, pp. 141 -1418, 2007.

[29] C. Damm, M. Schussler, J. Freese, R. lakoby, "Artificial Line Phase Shifter with Separately Tunable Phase and Line Impedance", Proc. 36th Eur. Microwave Conf., pp. 423-426, Sept. 2006.

[30] D. Kuylenstierna, E. Ash, A. Vorobiev, T. Itoh, S. Gevorgian", X-band Left Handed Phase Shifter Using Thin Film $YBa_2Cu_3O_7/SrTiO_3$ Ferroelectric Varactors," Proc. of 36th Eur. Microwave Conf:, pp. 847-8 0, Sept. 2006.

[31] Kambe, R., et al., "MCM Substrate with High Capacitance," IEEE Trans. Components, Packaging and Manufacturing Tech.—Part B, vol. 18, pp. 23–27, Feb. 199 .

[32] O. G. Vendik et al. , "Microwave losses in incipient ferroelectrics as functions of the temperature and the biasing field," J. Appl. Phys., vol. 84, no. 2, pp. 993-998, 1998.

[33] I. B. Vendik et al., "Commutation Quality Factor of Two-Sate Switchable Devices," IEEE Trans. Microwave Theory Tech., vol. 48, pp. 802-808, May 2000.

[34] C.-J. Wu, "Simulations of microwave characteristics of high-temperature superconducting microstrip lines by using an empirical two-fluid model," IEEE Trans. Appl. Superconduct. , vol. 12, no. 2, pp. 1776-1783, June 2002.

[35] A. Miranda, Joseph D. Warner, high-temperature superconductor and ferroelectric thin films for microwave applications, NASA Glenn Research Center, communication technology division, Cleveland, Ohio, USA.

[36] K. C. Gupta, R. Garg, I. Bahl, and P. Bhartia, Microstrip Lines and Slotlines, 2nd Ed. Artech House, 1996.

[37] S. Gevorgian, L. J. P. LinnCr, and E. L. Kollberg, "CAD Models for Shielded Multilayered CPW," IEEE Transactions on Microwave Theory and Techniques, vol. 43, no. 4, April 199.

Chapter 8

Paramagnetic Transitions Ions as Structural Modifiers in Ferroelectrics

Veronica Lucero Villegas Rueda

Abstract

The science of ferroelectric materials has long known that transition metal atom and/or rear earth atom substitution in the composition of a ferroelectric material can produce substantial structural and electric dipole changes and ferroelectric behavior. The focus is on first neighbor changes, symmetry, very tiny atomic displacements, hence magnitudes of electric polarization, charge changes, and mechanical-tensile change of parameters. The transition atom used for the substitution can, or, cannot be paramagnetic. When it is paramagnetic as is the case with Cr^{3+}, Mn^{2+} and so forth, there emerges an advantage for its experimental characterization at atomic level. Electron Paramagnetic Resonance (EPR) allows the identification of its location within the structure and the number and nature of its neighbors. The presence of crystal fields, symmetry and distortions of the first coordination sphere can also be determined. Here, we describe how a set of EPR spectra is analyzed to extract such atomic information.

Keywords: paramagnetic transition ions, ferroelectrics, PbTiO (Cr), octahedral symmetry, electron paramagnetic resonance, crystal field

1. Introduction

The ferroelectric materials are very important for technological applications in general, like sensor and actuators they are important part of electronic devices. The temperature, pressure, electric and electromagnetic field nonlinear response of ferroelectrics make them ideals like active elements due to pyroelectric and piezoelectric effects [1–4]. Ferroelectrics are used like ultrasonic generator; high voltage transforms and accelerators [1–4], also, they are using in optical components, piezoelectric transducers, and pyroelectric sensors [1–4]. By fabrication ferroelectric materials born with electric dipole different to zero, $P \neq 0$; this electric property is by asymmetry crystalline center structure. The **Figure 1** shows two crystalline structures, first is a symmetry perovskite structure with $P = 0$, and second is an asymmetry perovskite structure with $P \neq 0$ typical to lead titanate ($PbTiO_3$) [5]. The lead titanate ($PbTiO_3$) is used intensively in electronic devices [5–7]. The $PbTiO3$ ceramic had perovskite structure (tetragonal structure) with ratio $c/a = 1.064$, where $c = 4.154\text{Å}$, $a = 3.899\text{Å}$, the **Figure 1a** shows the lead ion (Pb) occupies places A, the titanium ion (Ti) occupies the places B and the oxygen ion is on the parallelepiped faces occupies the places O [8].

IntechOpen

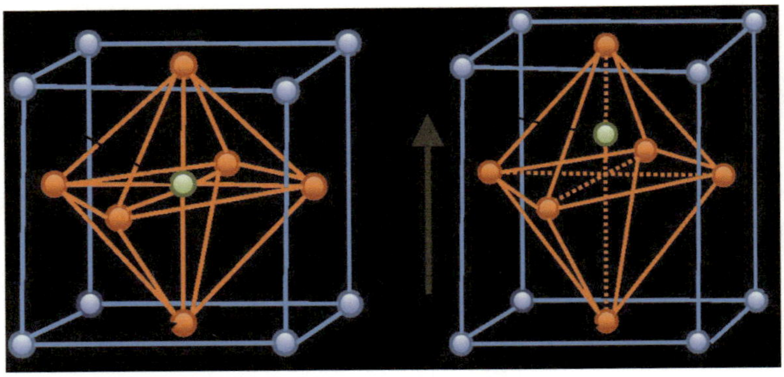

Figure 1.
The perovskite structure for $PbTiO_3$ a) Centre symmetry and b) non Centre with ratio $c/a = 1.064$, where $c = 4.154$Å, $a = 3.899$Å.

The ceramic fragility is reduced for inserting rare earths or transitions metal atoms [6–8]. The partial or total substitution of Pb or Ti produces modified compounds with new electrical polarization and modifies their structure depending of doping atoms and the interactions material [8]. The changes in the structure could be small or big depending to the kind of dopand and the percentage of them [1–8]. If the dopand is a paramagnetic ion it could be detected by paramagnetic resonance (EPR) and it is capable to sense structure changes principally by spin-orbit interactions [9–11].

The paramagnetic resonance technique (EPR) detects the spin-orbital magnetic interaction in doped ferroelectrics when the dopant is a rare earth with a paramagnetic ion. If dopant is chromium (Cr), it can give a paramagnetic state, in this case the electron paramagnetic resonance EPR technique is capable to detect paramagnetic ions, and the EPR-technique is high sensitivity to spin interactions due crystalline structure, nuclear interaction (hyperfine), or anisotropies (orientations) [9–11]. The nondestructive EPR-technique is applied to organic and inorganic molecules, ions, and atoms that have unpaired electrons that have information about oxidation state and spin state due to electron unpaired spin [9–11]. The EPR technique use microwave energy to induce resonances or electron transitions between electronic energy levels separated by applying magnetic field for paramagnetic compounds or substances. The EPR typical spectra for Cr, Fe, Mn, Co, Ni, Cu, V, etc. [11–15] contained information about the g-factor, the A-hyperfine couple and crystalline field factors, D and E.

2. Atoms into magnetic field $\vec{H}$

The energy interaction between magnetic moment $\vec{\mu}$ and extern magnetic field $\vec{H}$ is given by $W = -\vec{\mu} \bullet \vec{H}$. The electronic orbital movement is the cause of atomic momentum for the atom $\vec{\mu}$, it is proportional to angular momentum $\vec{L}$ given by $\vec{\mu}_L = \left(-\frac{e}{2mc}\right)\vec{L}$, where $\vec{L}$ is the orbital angular momentum, e is the electron charge, m is the electron mass, and c is the speed of light. The gyromagnetic ratio is defined by $\gamma \equiv \frac{e}{2mc} = 1.7 \times 10^7 rad\, gauss^{-1}$, so the atomic momentum is given by $\vec{\mu}_L = \gamma\vec{L}$ [9].

Otherwise, the intrinsic magnetic momentum for electron is $\vec{\mu}_S = 2\left(-\frac{e}{2mc}\right)\vec{S}$, where $\vec{S}$ is the spin vector associated to electron spin. The total magnetic moment is given by the spin and orbital momentum addition $\vec{\mu} = \vec{\mu}_L + \vec{\mu}_S$. If the total magnetic moment is placed into magnetic field $\vec{H}$, the total magnetic energy is given by $W = -\vec{\mu} \bullet \vec{H} = -\mu H \cos\theta$, where θ is the angle between vectors $\vec{\mu}$ and $\vec{H}$, the angle could varies continually in the classical description, however, in the quantum mechanics description the variation is quantized with $2J + 1$ orientations, where J is the quantum number for the total angular momentum given by $\sqrt{J(J+1)}\hbar$ [9, 12]. The allowed projections J, when the system is quantized along the magnetic field direction, are given by $m_j\hbar$, where m_j is the quantum magnetic number with values from $J, J-1, \dots, -J$ [15].

The simplest case is just only the spin electronic momentum (atoms in base state $^2S_{1/2}$) with $m_S = S, S-1, \dots, -S$, with S the total electronic spin, and the projections are $|\vec{\mu}_S| = 2(-\frac{eh}{4\pi mc})m_S$ where for $\vec{S}$ operator the eigenvalue $m_S\hbar$ was substituted. The quantity $\frac{eh}{4\pi mc} \equiv \beta_e = 9.2741 \times 10^{-21} erg/gauss$ is called Bohr's magneton, then $|\vec{\mu}_S| = \beta_e m_S$ and the energy values allowed for the atom placed into a magnetic field $\vec{H}$, are $E_{m_S} = -\mu_S H = 2\beta_e m_S H$ (Zeeman's energy). For an electron isolate quantum

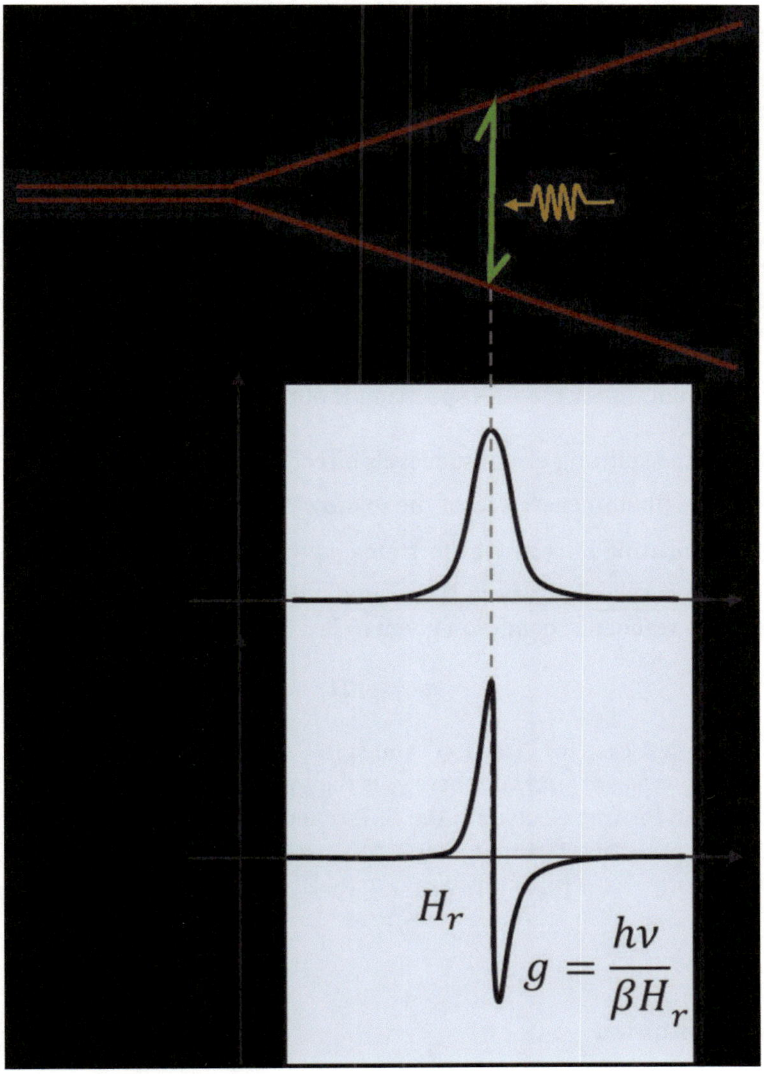

Figure 2.
EPR scheme for system with S = 1/2. The Zeeman's effect for S = 1/2 split the degenerate state energy into W_1 and W_2 and the microwave photon hν provide the energy for the transition between them [14].

electrodynamic correction is necessary [9–15] replacing the number 2 for the $g_e =$ 2.00023. In the case for isolate spin $S = 1/2$, the $2S + 1$ energy levels are $g_e\beta_e H$ equally spaced, the **Figure 2** shows the case for $S = 1/2$ [9–15].

For degenerated orbital state ($L \neq 0$) where exist and important L and S coupling (Russell-Saunders coupled) [9–15] there are $J = L + S, L + S - 1, \ldots, |L - S|$, and for each J there are $m_J = J, J - 1, \ldots, -J$ values, so the energy for each degenerated state is $E_{m_J} = g_J\beta_e m_J H$, where $g_J = 1 + \frac{J(J+1)+S(S+1)-L(L+1)}{2J(J+1)}$ is the splitting Landé factor. For example, the state $^2P_{1/2}$ have $S = 1/2, L = 1, J = 1/2$ and then $g = 3/2$. Of course, when $L = 0$ the $g_J = g_e$ [9].

3. The basic principle of the electron paramagnetic resonance (EPR) spectroscopy

The splitting of the degenerate energy levels applying magnetic field is the principle of the paramagnetic resonance spectroscopy useful for study the paramagnetic materials with total electronic spin $\vec{S} \neq 0$. The splitting of the electronic energy levels occurs when a magnetic field $\vec{H}$ is applied, this phenomenon is called Zeeman's effect. In the experiment an oscillating microwave field is applied to give energy to the electrons so the electronic transition of the electrons can occur; these electrons are arranged in the electronic levels according to Hund's rule. From quantum mechanics point of view, when the microwave energy photon $h\nu$ is equal to the energy difference between Zeeman levels there are electrons transitions from one low electronic energy level (W_1) to other with high energy (W_2); the energy difference between this electronic energy levels is $\Delta W = W_2 - W_1 = h\nu$, according to quantum mechanics selection rules [10, 12]. When $\Delta W = h\nu$ occurs, there are peaks of microwave energy absorptions observed in the EPR spectrum. Experimentally a high resolution for the EPR spectrum is obtained taken the first derivative of absorption [9–15].

In general, the splitting electronic levels effect is written tensorial form by $\mu_\beta \vec{S} \bullet \overleftrightarrow{g} \bullet \vec{H}$. The photon energy hv of the microwaves is provided to the system applying an oscillating microwave electromagnetic field $\vec{H}_1$ at frequency ν_0; the $\vec{H}_1(t)$ is applied perpendicularly to $\vec{H}_0$ magnetic field, and the magnitude of H_0 is varied until the resonance condition given in Eq. (1) be satisfied.

$$h\nu = g\beta H \tag{1}$$

For the simplest case for "free ion" and spin $S = 1/2$ system the Eq. (1) is reduced to $\Delta W = h\nu_0 = g_e\beta_e H_0$, where g_e is the gyromagnetic ratio and β_e is the Bohr's magneton for free electron e; the EPR scheme is shown in **Figure 2**. The constants value $g_e = 2.0023$ and $\beta_e = 9.2741 \times 10^{-21} erg/gauss$, the Planck's constant is $h = 6.6262 \times 10^{-27} ergs$ [9, 15]. In general, these energy levels are described by tensorial treatment.

4. Spin Hamiltonian

For a paramagnetic ion, the energy levels are quantized. This energy levels are eigenvalues of the spin Hamiltonian operator which is the representation of total electronic energy for the ion or system. Usually the lowest energy levels states are populated at ordinary temperatures of $300K$ ($\approx 200 cm^{-1}$) and this is the group of levels of the ground state. The ground state is the most of the times the only involved in the resonance experiment, it is to say, the transition are induced between this energy levels of ground state under microwave excitation. The energy of each level depends of the ion properties like electric charge, mass, atomic number, etc. and the energy level depends too to the crystalline field effect and the external magnetic field applied along with appropriate nuclear interactions [9, 12, 15].

The EPR results are interpreted by spin Hamiltonian that describes the system and the interactions mentioned above. The Hamiltonian is given in general by Eq. (2) and Eq. (3) [9–12].

$$\hat{H} = \beta\vec{S} \bullet \overleftrightarrow{g} \bullet \vec{H} + \vec{S} \bullet \overleftrightarrow{A} \bullet \vec{I} + \vec{S} \bullet \overleftrightarrow{D} \bullet \vec{S} \tag{2}$$

$$\hat{H} = g\beta\vec{S} \bullet \vec{H} + A\vec{S} \bullet \vec{I} + D\left[S_x^2 - \frac{1}{3}S(S+1)\right] + E\left(S_x^2 - S_y^2\right) \tag{3}$$

The first term is by Zeeman electronic interaction, the second term is the representation of hyperfine interaction, the third and fourth terms are due to the crystalline field, where $\overleftrightarrow{g}$, $\overleftrightarrow{A}$ and $\overleftrightarrow{D}$ are the spectroscopy, hyperfine interaction and crystalline field third order tensors respectively.

The magnetic interaction is naturally anisotropic, and the tensors are used to describe it, like the magnetic moment for each electron $\overleftrightarrow{\mu}$. The anisotropy property for magnetic moment $\overleftrightarrow{\mu}$ is measured by spectroscopy factor $\overleftrightarrow{g}$.

The solutions for the Hamiltonian are compared with measures of g parameters in the spectrum EPR and other paramagnetic parameters in the system [9–12].

4.1 Zeeman electronic term

The general expression for Zeeman interaction between external magnetic field $\vec{H}_0$ and the electronic spin $\vec{S}$ is given in Eq. (4) and it is rewritten in terms of matrix in Eq. (5) [9–15].

$$\hat{H}_{Ze} = \beta\vec{S} \bullet \overleftrightarrow{g} \bullet \vec{H}_0 \tag{4}$$

$$\hat{H}_{Ze} = \beta\begin{pmatrix} S_x & S_y & S_z \end{pmatrix} \begin{pmatrix} g_{xx} & g_{xy} & g_{xz} \\ g_{yx} & g_{yy} & g_{yz} \\ g_{zx} & g_{zy} & g_{zz} \end{pmatrix} \begin{pmatrix} H_x \\ H_y \\ H_z \end{pmatrix} \tag{5}$$

Where H_x, H_y, H_z, S_x, S_y, S_z, are the three scalar components for external magnetic field $\vec{H}_0$ and $\vec{S}$ in a fixed Cartesians coordinates x, y and z in the molecule.

Many times is found that tensor $\overleftrightarrow{g}$ is a symmetric matrix, which could be diagonalized through appropriate transformation [9, 16] $M\overleftrightarrow{g}M^{-1} = \overleftrightarrow{g}_{diagonal}$. This transformation corresponds to axes reorientation and the matrix M redefine the orientation for new principal axes respecting to previous axes. After to diagonalized the Zeeman's Hamiltonian writes like Eq. (6).

$$\hat{H}_{Ze} = \beta\left(g_{xx}H_xS_x + g_{yy}H_yS_y + g_{zz}H_zS_z\right) \tag{6}$$

The $\overleftrightarrow{g}$ components g_{xx}, g_{yy} and g_{zz} measured the contribution of magnetic moment along principal direction xx, yy and zz of magnetic field. There is spherical symmetry for the electron $\overleftrightarrow{\mu}(g)$ when $g_{xx} = g_{yy} = g_{zz}$.

The Hydrogen atom have spherical symmetry, the spin Hamiltonian have a g isotropic factor for an electron and isotropic hyperfine interaction, A, between electron and nucleus. In the most molecules these quantities vary with applied magnetic field direction and the spin Hamiltonian is anisotropic [9–16].

4.2 Axial symmetry

If the $\overleftrightarrow{g}$ tensor is anisotropic, there is the axial symmetry when $g_{xx} = g_{yy} \neq g_{zz}$. Usually the EPR bibliography writes $g_{xx} = g_{yy} = g_{\perp}$ and $g_{zz} = g_{\parallel}$ [9–12].

Suppose that magnetic field $\vec{H}$ is applied with θ angle respect to $z-$ axis. Rewriting the components, $H \cos\theta$ is parallel to z and $H \sin\theta$ is parallel to x, them the Zeeman $\hat{H}_{Ze}$ writes like Eq. (7) [9].

$$\hat{H}_{Ze} = \beta H \left(g_{\parallel} S_z \cos\theta + g_{\perp} S_x \sin\theta \right) \tag{7}$$

Where $S_x = \frac{1}{2}(S_+ + S_-)$ written in terms of created and annihilated spin operators. With low symmetry and magnetic field random oriented, using the director cosines l, m, n with respect x, y and z axis, then the spin Hamiltonian writes like Eq. (8).

$$\hat{H}_{Ze} = \beta H \left(g_{xx} l S_x + g_{yy} m S_y + g_{zz} n S_z \right) \tag{8}$$

This correspond to rhombic symmetry $g_{xx} \neq g_{yy} \neq g_{zz}$.

For the axial asymmetry g factor is dependent of angle θ by $g(\theta)^2 = g_{\perp}^2 \sin\theta + g_{\parallel}^2 \cos\theta$, for an axial g matrix with $g_{\parallel} > g_{\perp}$, the line shape of the corresponding EPR spectrum are drawn in **Figure 3**, assuming a large number of paramagnetic systems with random orientation of their $\overleftrightarrow{g}$ ellipsoids with respect to the static magnetic field $\vec{H}$ [9]. This situation is typical for a powder sample that contain all possible direction $g-$ ellipsoids. For a given magnetic strength H, all spins fulfilling the resonance condition $g(\theta) = h\nu/\beta H$, i. e., all spins for which H makes an angle θ with the $z-$ axis of the $g-$ ellipsoid, contribute to the spectrum and are considered to form a spin packet. The extreme positions ($\theta = 0°$ and $\theta = 90°$) of the powder spectrum are obtained by inserting $g_{\parallel}$ and $g_{\perp}$ into the resonance condition. If $g_{\parallel} > g_{\perp}$ the asymmetry line shape is mainly due to the fact that the number of the spin packets contributing to the spectrum is much larger in the xy-plane than the along the $z-$ axis. If $g_{\parallel} < g_{\perp}$ the asymmetry line shape is mainly due to the fact that the number of the spin packets contributing to the spectrum is much larger along the $z-$ axis than in the xy-plane. Also, by definition the $g_{iso} = \sqrt{\frac{1}{3}\left[g_{xx}^2 + g_{yy}^2 + g_{zz}^2\right]}$ [9–17].

4.3 Hyperfine interaction term

The hyperfine interaction is the interaction between decoupled electrons $(\vec{S})$ and nuclear magnetic moments $(\vec{I})$ of several neighboring nucleus. In the EPR spectra the hyperfine split is due to interaction between electronic spin and nuclear spin, it causes splitting Zeeman levels. Each Zeeman level is splitting $2I + 1$ times, where I is the nuclear spin [12, 16]. In tensorial form the Hamiltonian spin term is given by $\vec{S} \bullet \overleftrightarrow{A} \bullet \vec{I}$. Similarly to g-factor, the A -hyperfine factor give the magnitude

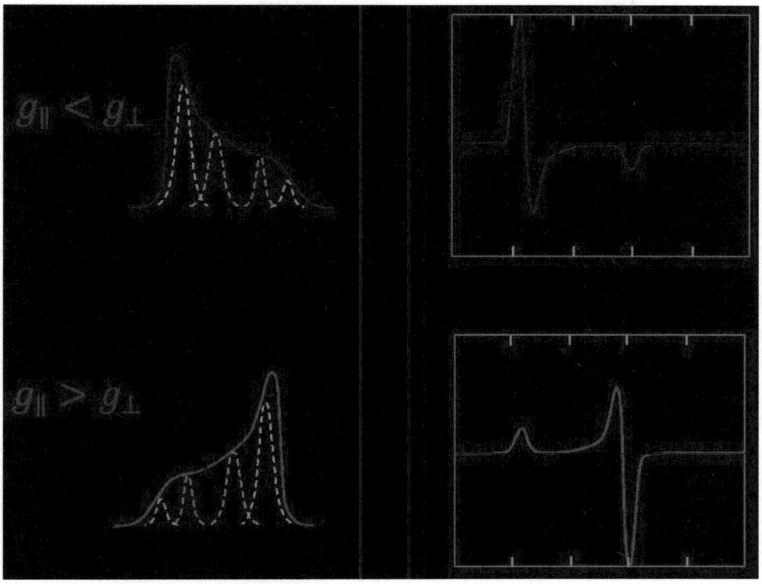

Figure 3.
Microwave absorption spectra and their first derivate for axial symmetry (A) $g_{\|} > g_{\perp}$ and (B) $g_{\|} < g_{\perp}$. Both cases show all contributions for all directions contributions of $g(\theta) = h\nu/\beta H$ in a powder samples [9–16].

of hyperfine interaction, in general it is an anisotropic tensor. There are two types of hyperfine interaction [9–15].

The first interaction is the classic interaction between $\vec{\mu}_S$ and $\vec{\mu}_I$ dipoles separated a $\vec{r}$ distance given by Eq. (9) [9, 12].

$$E_{dip} = \frac{\vec{\mu}_S \bullet \vec{\mu}_I}{r^3} - \frac{3\left(\vec{\mu}_S \bullet \vec{r}\right)\left(\vec{\mu}_I \bullet \vec{r}\right)}{r^5} \tag{9}$$

For correspondence principle, the quantum Hamiltonian for this interaction is given in Eq. (10).

$$\hat{H}_{dip} = g_e\beta_e g_n\beta_n \left[\frac{\vec{S} \bullet \vec{I}}{r^3} + \frac{3\left(\vec{I} \bullet \vec{r}\right)\left(\vec{S} \bullet \vec{r}\right)}{r^5}\right] \tag{10}$$

The second interaction is no classic interaction and comes from to the probability different of zero for found an electron in the nuclear region $(0 < r < a_0)$, where a_0 is the Bohr's radii, i.e., it is proportional to the square electronic function valuated in the nucleus. Fermi proof that this interaction is isotropy and is called contact interaction or Fermi's interactions, given by Eq. (11) [10–17]. Here $\Psi(0)$ is the electronic wave function valued in the nucleus.

$$\hat{H} = \left(\frac{8\pi}{3}\right)\vec{\mu}_S \bullet \vec{\mu}_n |\Psi(0)|^2 \tag{11}$$

If the molecule have one or more neighboring nuclei to the uncoupled magnetic dipolar momentum, it turns out split hyperfine of energy magnetic levels of the decoupled electron (even without external magnetic field applied) due to interaction of each nucleus with the electronic magnetic momentum.

When the conditions are favorable the hyperfine interaction could be measured like in the case of the hyperfine interaction splitting of an octahedral manganese(II) complex with spin $S = 5/2$ and nuclear spin $I = 5/2$ and the corresponding EPR spectrum are shown in the Drago's book at Figure 13.10 [9].

In the simplest case, the energy levels of a system with one unpaired electron and one nucleus with $I = 1/2$ are show in **Figure 4(A)** with sufficiently high fixed magnetic field $\vec{H}$. The dashed line would be the transition corresponding to $\Delta W = h\nu = g\beta H$ in the absence of hyperfine (A_0), like is show in **Figure 2**. The solid lines marked ΔW_1 and ΔW_2 correspond to the allowed EPR transitions with the hyperfine coupling operative. To first order, $h\nu = g\beta H \pm \frac{1}{2}A_0$, where A_0 is the isotropic hyperfine coupling constant. In **Figure 4(B)** are shown the hyperfine splitting as a function of an applied magnetic field. The dashed line corresponds to the transitions in the case of $A_0 = 0$. The solid lines ΔW_1 and ΔW_2 refer to transitions induced by a constant microwave quantum $h\nu$ of the same energy as for the transition ΔW. Here the resonant-field values corresponding to these two transitions are, to first order, given by $H = h\nu/g\beta \pm 1/2(g_e/g)a_0$ (in mT) is the hyperfine spplitting constant given approximately by $H_{\Delta W_1} - H_{\Delta W_2}$. Note that these diagrams are specifics to a nucleus with positive g_n and A_0 values, such as hydrogen atom [9, 12]. The **Figure 4 (C)** shows the typical EPR spectrum and the $g-value$ approximate positions for the

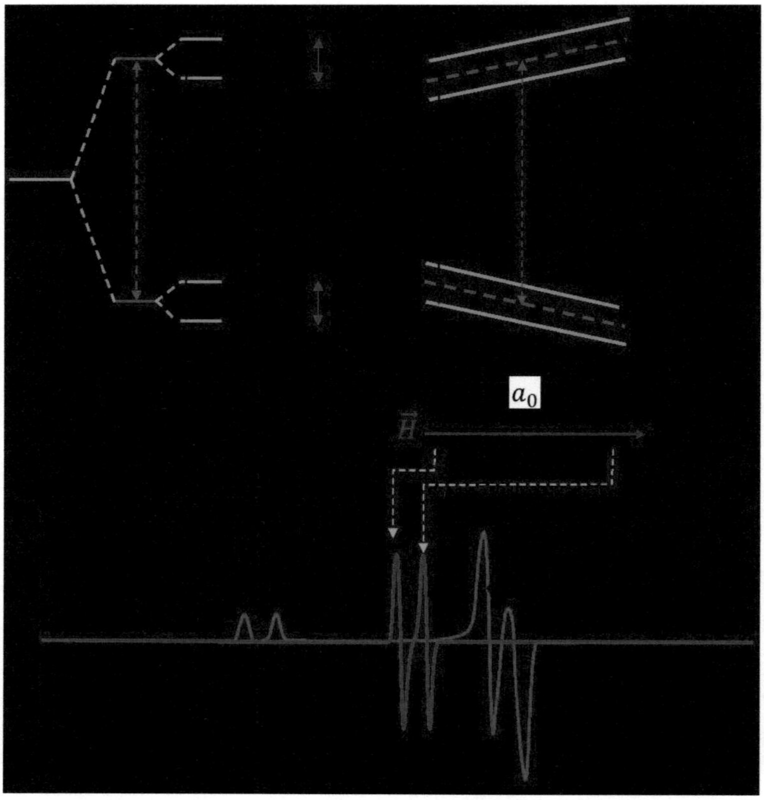

Figure 4.
(A) At a sufficiently high fixed magnetic field $\vec{\mathbf{H}}$. *The dashed line would be the transition corresponding to* $\Delta W = h\nu = g\beta H$ *in the absence of hyperfine* (A_0). *The solid lines marked* ΔW_1 *and* ΔW_2 *correspond to the allowed EPR transitions with the hyperfine coupling operative. To first order,* $\Delta W = h\nu = g\beta H \pm \frac{1}{2}A_0$, *where* A_0 *is the isotropic hyperfine coupling constant. (B) As a function of an applied magnetic field. The dashed line corresponds to the transitions in the hypothetical case of* $A_0 = 0$. *The solid lines* ΔW_1 *and* ΔW_2 *refer to transitions induced by a constant microwave quantum* $h\nu$ *of the same energy as for the transition* ΔW. *Here the resonant-field values corresponding to these two transitions are, to first order, given by* $H = h\nu/g\beta \pm 1/2(g_e/g)a_0$ *(in mT) is the hyperfine spplitting constant. (C) Shows the typical EPR spectrum and the* $g-$ ***value*** *approximate positions for the case described in (A) and (B) with* a_0 *hyperfine splitting, and additionally shows the hyperfine splitting for and axial symmetry EPR signal with* $A_{\parallel}$ *and* $A_{\perp}$ *splitting factor than indicating how hyperfine interactions is affecting the energy levels transitions [11, 12].*

case described in (A) and (B) with a_0 hyperfine splitting, and additionally shows the hyperfine splitting for and axial symmetry EPR signal with $A_{\parallel}$ and $A_{\perp}$ splitting factor than indicating how hyperfine interactions is affecting the energy levels transitions [11, 12].

4.4 Crystalline field term

Other observable interaction in EPR is the crystalline field [9–17], this interaction is represented by tensors given by $\vec{S} \bullet \overleftrightarrow{D} \bullet \vec{S}$. This is the interaction of electron spin with the electric field of the charges of the neighboring ions placed in specific symmetries. The expression for crystalline field is given by [9–12]:

$$D\left[S_x^2 - \frac{1}{3}S(S+1)\right] + E\left(S_x^2 - S_y^2\right) \tag{12}$$

The Zero-field splitting parameters D and E split the energies of m_s levels at zero applied magnetic field according to the magnitude of m_s for each level. Thus the D and E ensure the singularity of each energy gap between m_s levels under a non-zero magnetic field. The magnitudes of D and E are dependent on the ligant field theory (or from the Crystal Field Theory point view), and therefore are easily tunable by coordination geometry [12, 18]. Even the $g-$ factor could give in terms of crystal splitting factor Δ [11, 12, 18].

The **Figure 5** shows how the splitting of electronic levels depends on the ion interaction with electric field of the neighboring charge ions placed in specific symmetries [9–12].

The crystal field splitting energy $\boldsymbol{\Delta}$ is the energy of repulsion between electrons of the ligands and the central metal ion and their bounding in complex ions such as octahedral, square planar and tetrahedral structural symmetries [9, 12]. If the $\boldsymbol{\Delta}$ is greater than electron spin pairing energy the greater stability would be obtained if the fourth and fifth electrons get paired with the ones in lower level. If the $\boldsymbol{\Delta}$ is less than the pairing energy, greater stability is obtained by keeping the electrons unpaired. So, if $\boldsymbol{\Delta}$ is weak then the spin $\boldsymbol{S}$ is high and this yields a strongly paramagnetic complexes, and if $\boldsymbol{\Delta}$ is high then the spin S is weak and this yields low spin complexes and weakly paramagnetic or sometimes even diamagnetic. Through microwave excitation the electronic transition energy levels are possible when this obey the rules for allowed transitions [11, 12].

For example for octahedral symmetry, the tetragonal distortion could provide a high Δ for $\boldsymbol{d^3}$ electrons them could be arrangement in levels $\boldsymbol{B_{2g}}$ and $\boldsymbol{E_g}$ resulting spin S = 1/2, **Figure 6(A)**. For example for tetrahedral symmetry, the tetragonal distortion could provide a high Δ for $\boldsymbol{d^3}$ electrons them could be arrangement in levels $\boldsymbol{A_{1g}}$ and $\boldsymbol{B_{1g}}$ resulting spin S = 1/2, **Figure 6(B)**.

For the tetragonal symmetry, the degeneracy of the $\boldsymbol{E_g}$ ($\boldsymbol{d_{z^2}}$ and $\boldsymbol{d_{x^2-y^2}}$ orbitals) term is no affected by the spin orbit coupling and by Jahn-Teller theorem applies. The orbital degeneracy is lifted and the energy of the system lowered by a displacement of the ligands on the z-axis [11, 12]. An elongated or compressed of the coordination tetrahedron (or tetragonal distortion) leads to the energy level scheme shown in **Figure 6(B)** with unpaired electron in the $\boldsymbol{d_{x^2-y^2}}$ orbital. The measure of EPR spectra is limited to the Zeeman splitting imposed by an external field on the unpaired electron in the non-degenerated $\boldsymbol{d_{x^2-y^2}}$ orbital and one could think that the spin system can now be described by the spin Hamiltonian with S = 1/2 since the ground state is non-degenerate and has only associated spin angular momentum [9–15].

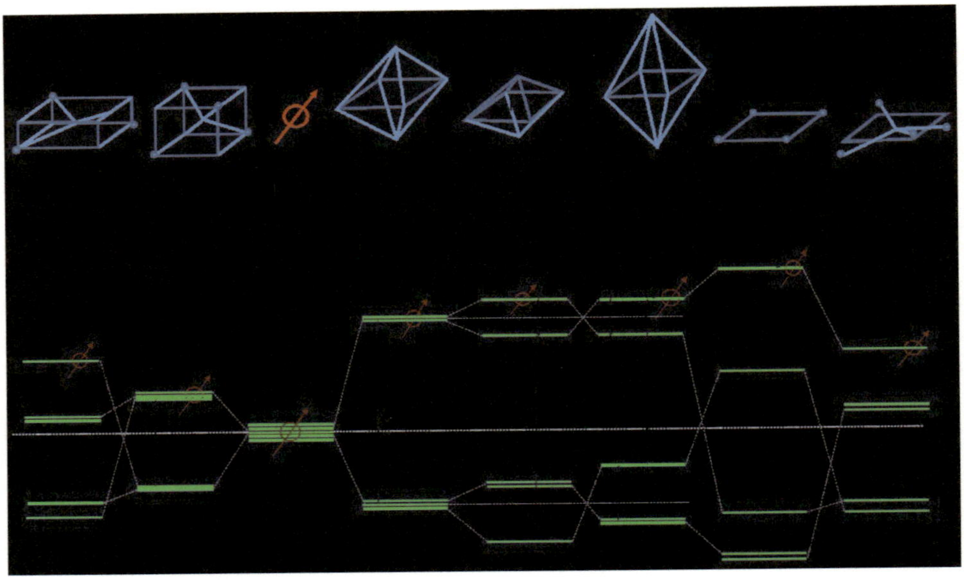

Figure 5.
Crystalline field interaction for a 3d electron with neighboring placed in specific symmetries, below are the corresponding splits electronic levels. The splitting levels have crystal field splitting energy Δ or in terms of D and E zero-field splitting parameters [9–17].

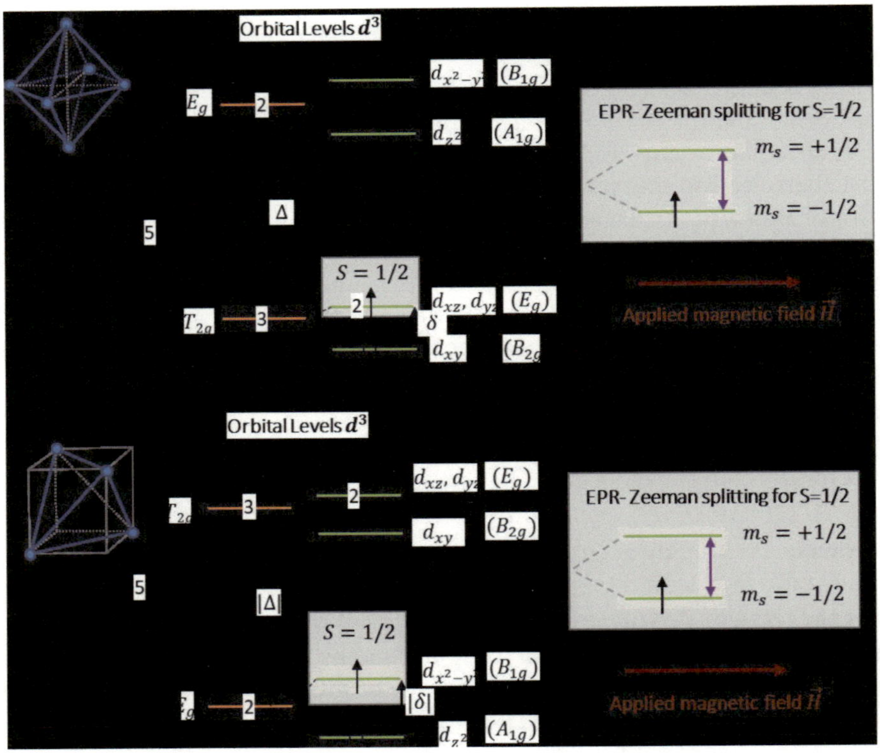

Figure 6.
Splittings and degeneracies of orbital levels d^1 or d^6 ions in two types of electric field caused by negative charges for (A) octahedral field (Δ > 0) plus tetragonal distortion and (B) tetrahedral field (Δ < 0) plus tetragonal distortion. For d^4 and d^9 ions applieds to octahedral and tetrahedral fields. Shifting of the center of gravity of the set of levels is ignored [9–15]. If **Δ** *is high, then the spin* **S** *is weak, and this yields low spin complexes and weakly paramagnetic. For example, for tetrahedral field, if* **Δ** *is high for* $\boldsymbol{d^3}$ *electrons them could be arrangement in levels* $\boldsymbol{A_{1g}}$ *and* $\boldsymbol{B_{1g}}$ *yield a spin S = 1/2.*

5. Characterization ceramic

The $Pb_{0.95}Sr_{0.05}(Zr_{0.53}Ti_{0.47})O_3 + x\%wtCr_2O_3$ is a ferroelectric ceramic powder, and it is possible to characterized by x-ray and electric measures [8, 16]. The EPR measures were taking at Biophysics and Magnetic Measure National Polytechnic Institute Laboratory using an EPR JEOL JES-RE3X spectrometer, **Figure 7**.

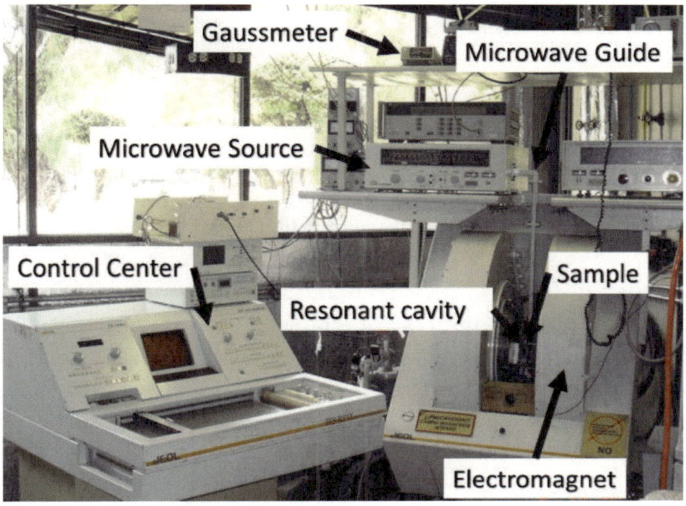

Figure 7.
EPR JEOL spectrometer [23].

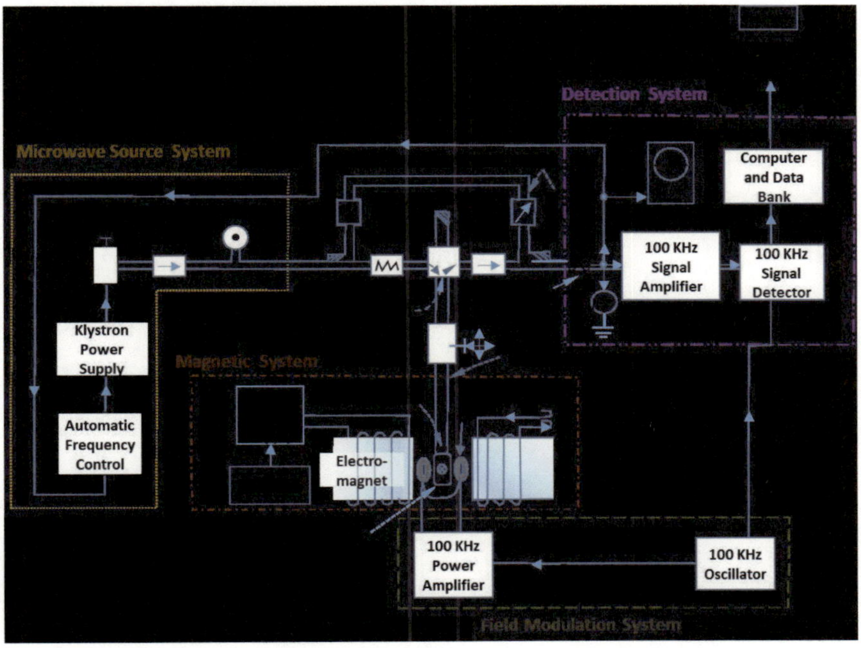

Figure 8.
General diagram for paramagnetic resonance spectrometer (EPR) that consist of magnetic system (electromagnet, modulation coils, magnet power supply and field scan drive), field modulation system (power amplifier, 100 kHz oscillator), cavity system (resonant cavity with sample, iris and the T-magic system), microwave source system (klystron power supply, automatic frequency control, etc.) and detection system (detector crystal, 100 kHz signal detector, 100 kHz signal amplifier, oscilloscope, computer and data bank; and display of the spectrum) [12].

The temperature measures were at 300K and 77K. The power potency was varied from 1 mW until 40 mW [17–23], at 9.45GHz microwave frequency at X-band. The lead zirconate titanate was doped with five percentages of Cr of 0%, 1%, 2%, 4% and 5%, we called the samples 0, 1, 2, 4 and 5 respectively.

The spectrometer is connected to workstation ES-PRIT to HP-9000 computer with a converter analogic digital target (A/D). The programing package performs acquisition, procession and simulation data [23, 24]. The block diagram typical X-band EPR spectrometer employing 100KHz phase sensitive detection is shows in **Figure 8** [12, 23].

6. Results and discussions

The X-rays and electric measured was published by F. Calderon, and Yañez, et al. [8], the x-rays spectra are show in **Figure 9** for samples 4 and 5. The X-rays program fixed the $Pb(Zr_{0.44}Ti_{0.56})O_3$ compound and this do not detect secondary phases. The quantity of chromium is not enough for be detected by X-rays measures, because the limit detection is 5% of the element. About electric measured, when the chromium concentration varies from sample 1 to 5, the Curie's temperature value increases. The Cr substitution causes a tetragonal distortion field to the tetrahedral symmetry structure causes energy splitting effect by zero splitting field [12]. Due to the Cr^{3+} quantity is more than Ti^{4+} or Zr^{4+} quantities, the system tetragonality be modified and the grain size is increased [8]. The Curie's temperature varies slightly, and the energy activation is below and above to the transition.

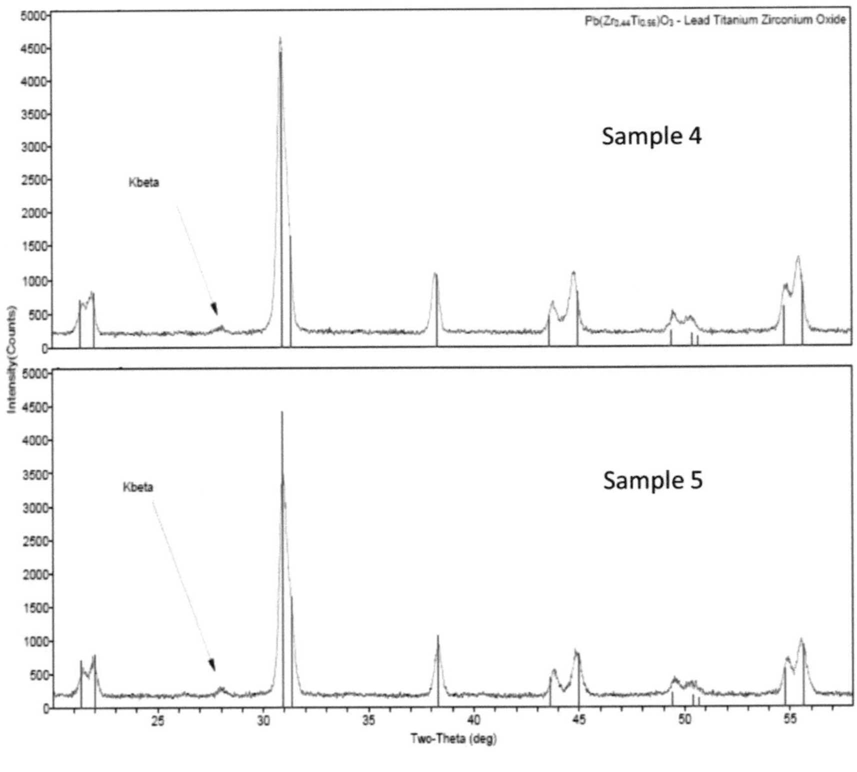

Figure 9.
X-rays for 4 and 5 samples that corresponds to perovskite structure or tetrahedral symmetry from Pb $(Zr_{0.44}Ti_{0.56})O_3$.

The presence of chromium determined by EPR guaranteed an oxygen or hold vacancy mechanism, also this effect was confirmed by Yañez, et al. [8].

6.1 EPR measured and results

The EPR spectrum were obtained for powder ferroelectric at 300 K. The EPR measures for the zero sample produce a spectrum with signals R and C, see **Figure 10**. The signal R had $g = 2.1295$ at 317 mT field. The signal C had several values $g_{\parallel} = 1.9194, g_{\perp} = 1.9355$ and $g_{iso} = 1.9301$. The intensity of signal C is low and this indicate that insipidus presence of the paramagnetic center. The signal corresponds to uniaxial local symmetry on the tetrahedral perovskite structure [12]. Even when the microwave power is increased there are no signals of saturation effects for all samples, this indicate that the paramagnetic center is stable and maintain their interaction neighbor stable.

The 1, 2 and 5 samples at 300 K and 77 K shows the same spectrum except the intensity increases from sample 2 to sample 5, the spectra for samples 2 and 5 are show in **Figures 11** and **12** respectively. For sample 1, the signal B had $g_{B\perp} = 1.9731$ and $g_{B\parallel} = 1.9441$; the signal B^* had $g_{B^*\perp} = 1.9352$ and $g_{B^*\parallel} = 1.9257$ spectroscopy parameters. For sample 2, there are two signals B and B^* too showed in **Figure 8**, for signal B was obtained $g_{B\perp} = 1.9720$ and $g_{B\parallel} = 1.9462$ at 345.20 mT and 350.31 mT respectively magnetic fields. For signal $B^* g_{B^*\perp} = 1.9360$ and $g_{B^*\parallel} = 1.9204$ at 352.14 mT and 355.00 mT respectively magnetic fields. When the Cr^{+3} substituted the Ti^{+4} ion, the tetragonality decreases [8].

The **Figure 13** shows the spectrum for sample 4 with the signals B and B^* too, but additionally shows the G, H and $DPPH$ signals. For signal B the $g_{B\perp} = 1.9762$ and $g_{B\parallel} = 1.9424$ at 344.74 mT and 350.71 mT respectively. For signal $B^* g_{B^*\perp} = 1.9384$ and $g_{B^*\parallel} = 1.9189$ at 351.43 mT and 355.00 mT respectively. The G signal had $g_G = 2.0755$ at 328.21 mT. The signal H had $g_H = 2.1449$ at 317.6 mT field. The signal for $DPPH$ had $g = 2.0036$ at 340 mT expected value for the EPR standard paramagnetic marker.

The analysis of EPR results start from chemical composition for ferroelectric $Pb_{0.95}Sr_{0.05}(Zr_{0.53}Ti_{0.47})O_3 + x\%wtCr_2O_3$ from the spin electronic arrangements for

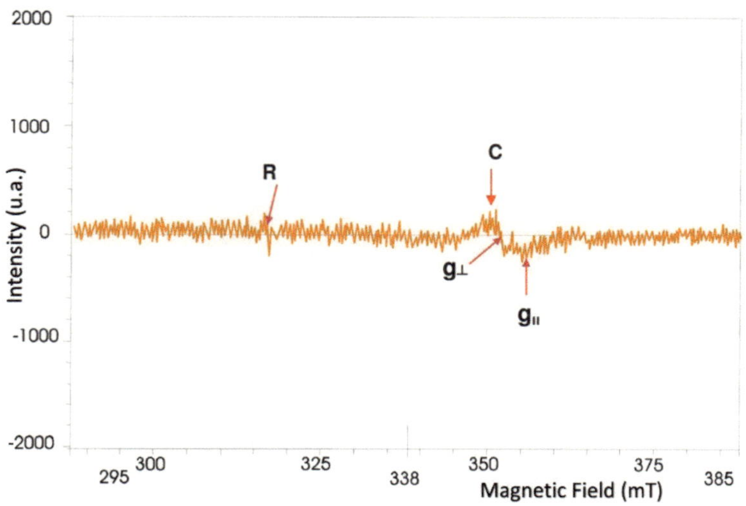

Figure 10.
Zero sample spectrum shows R and C signals.

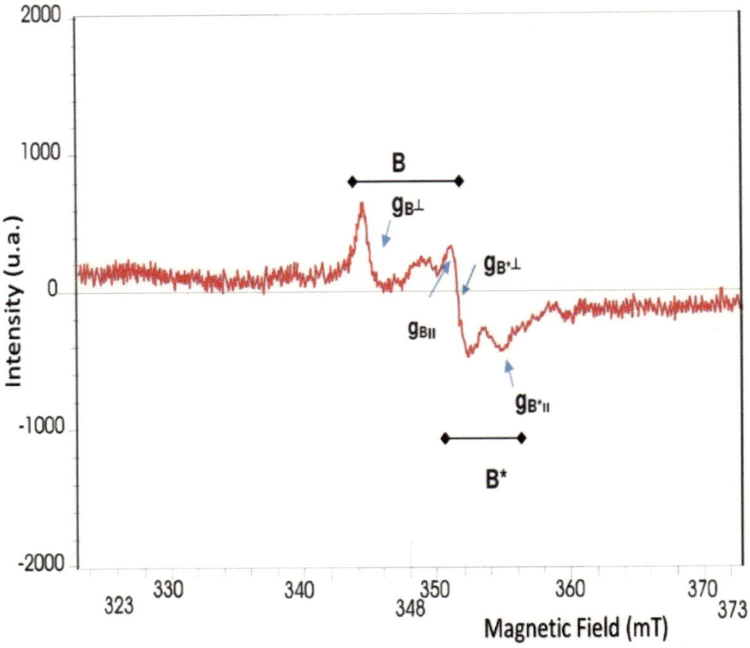

Figure 11.
EPR spectrum for sample 2. There are two signal groups B and B.*

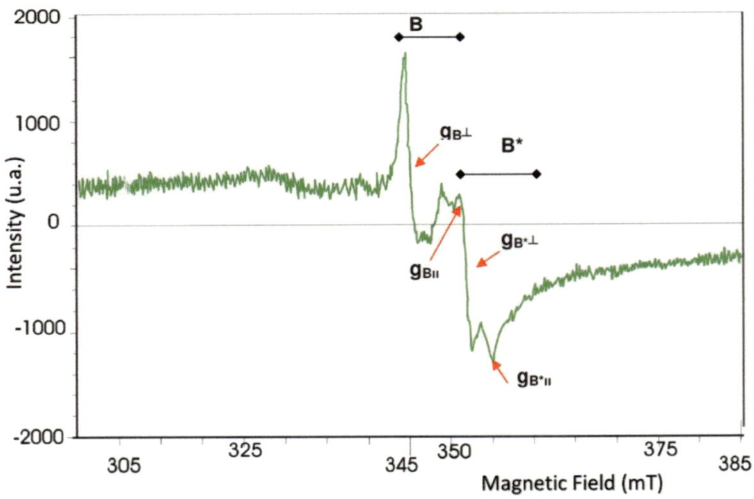

Figure 12.
EPR Spectrum for sample 5.

each element atom composition. The *Cr* had $1s^2\ 2s^2\ 2p^6\ 3s^2\ 3p^6\ 4s^2\ 3d^4$ electronic arrangement with +6, + 5, +3 and +2 oxidation states. The chromium compounds with Cr^{+6} oxidation state are no paramagnetic, because the $1s^2\ 2s^2\ 2p^6\ 3s^2\ 3p^6\ 4s^0\ 3d^0$ electronic configuration had no unpaired electron. The Cr^{+5} had $1s^2\ 2s^2\ 2p^6\ 3s^2\ 3p^6\ 4s^0\ 3d^1$ electronic configuration with unpaired electron and 5/2, 3/2 and ½ spin state corresponding to low, medium and high spin respectively. For Cr^{+3} the spin states could be ½ and 3/2, corresponding to weak

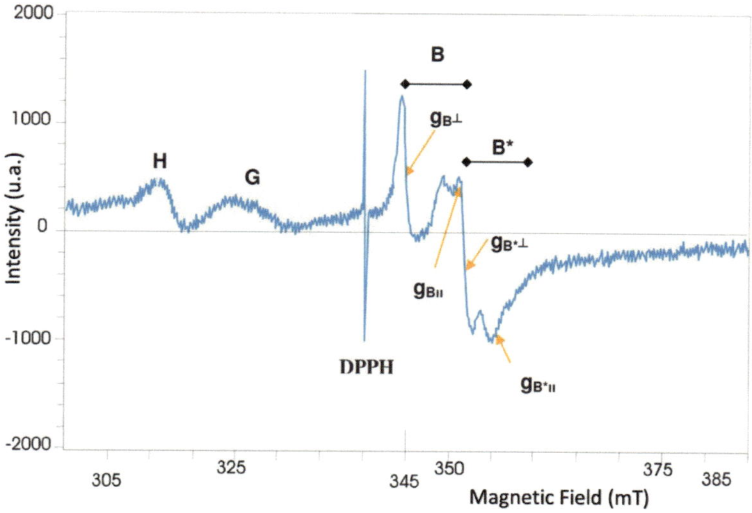

Figure 13.
EPR spectrum for sample 4.

and medium crystalline field split respectively. The Cr^{+2} is no paramagnetic by the $1s^2\ 2s^2\ 2p^6\ 3s^2\ 3p^6\ 4s^0\ 3d^4$ electronic configuration, i.e., had four paired electrons in d-orbital, this is because we have a octahedral crystalline structure into the perovskite structure that causes electronic levels split (Zero field splitting); this split have a high energy levels separation and the electrons unfollows Hund's rule, see **Figure 6(A)**, this causes that electrons stays at lowers energy levels, all this is because the experimental g-values are less than 2.00 for all spectrum, **Figure 10–13**. The other elements are not paramagnetic [19, 20, 24–26]. The titanium has 4+ the state oxidation (Ti^{4+}), and the lead state oxidation is +2 (Pb^{2+}), both elements have not uncoupled electrons and they cannot be detected by EPR. Only Cr^{3+} or Cr^{5+} signals are expected in the EPR spectrum.

From chemical composition $Pb_{0.95}Sr_{0.05}(Zr_{0.53}Ti_{0.47})O_3 + x\%wtCr_2O_3$ with $x =$ 0.0, 0.1, 0.2, 0.4 and 0.5, to sites B the Zr^{4+} is substitute for Ti^{4+} and to sites A the Sr^{2+} is substitute for Pb^{2+} in perovskite structure of the material [1–4, 25–27]. Additionally, in **Figure 14** shows how Cr is introduced to substitute Ti in sites B for the samples [8, 25–27]. The sample zero no contained Cr and it was taken like the control sample or EPR blank sample.

The **Figure 15** shows the EPR spectrum for sample 0 at 300 K. The R signal is at 317 mT and $g = 2.0134$. The C signal is axial $g_\perp = 1.965$ and $g_\parallel = 1.9181$; bout signals are small at noise level. Thus, the blank sample, whose chemical formula indicates that it should not give an EPR spectrum, shows the R resonance, which is

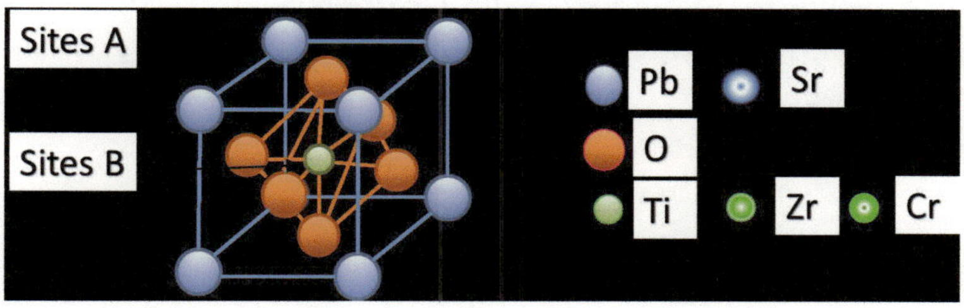

Figure 14.
Perovskite structure for ferroelectric $Pb_{0.95}Sr_{0.05}(Zr_{0.53}Ti_{0.47})O_3 + x\%wtCr_2O_3$ *with* $x = 0.0, 0.1, 0.2, 0.4$ *and* 0.5.

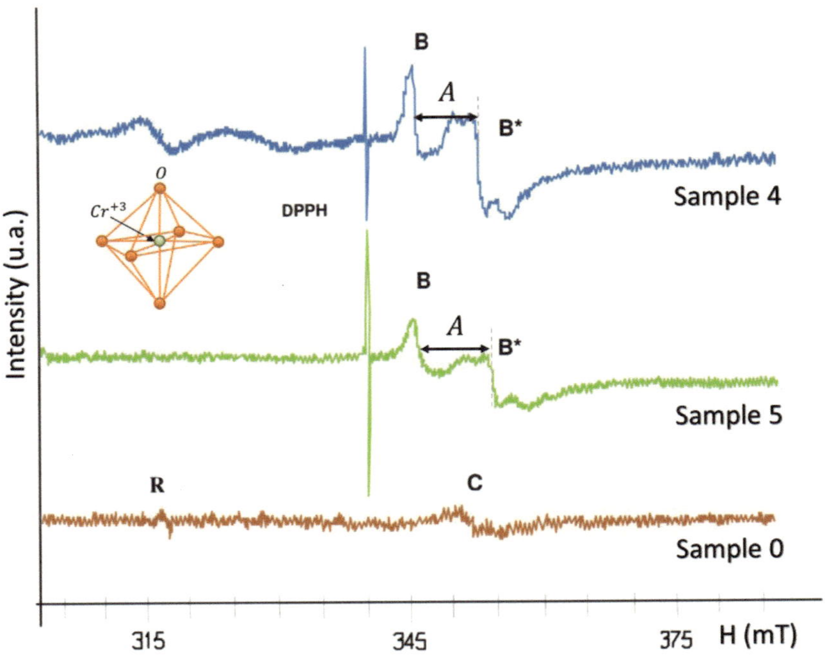

Figure 15.
EPR spectra for 1, 5 and 4 samples at 300 K. the two B and B^* oblate axial signals are typical by powder samples with $g_{\perp} > g_{\parallel}$ and hyperfine interaction $a_o = A$.

due to Fe^{3+}, and the *C* axial signal corresponds to *Cr* spectrum. When the temperature is low to 77 K, the *R* signal disappears and the *C* signal decreases. The *Cr* in this sample comes as manufacturing impurity. The 1, 2 and 5 samples show at 300 K and 77 K the same line shape spectrum, and the intensity increase from 1 to 2 and to 5. The increased intensity is due to percentage increase of *Cr* from 1% to 2% and 5% because the area under curve absorption EPR is proportional to the number of paramagnetic ions [9–12]. The typical EPR spectrum for the samples 0, 4 and 5 with *g* values less to 2.00 were obtained, see **Figure 15**. The *DPPH* signal is at left to the spectrum this means that *Cr* paramagnetic ions are d^3 with low spin ½ of Cr^{3+} [9–18]. This case is analogous for Mn^{4+}, $S = 1/2$ in the ferroelectric $Pb(Ti, Mn)O_3$ [27] that is isoelectronic with Cr^{3+}. The low spin value of $S = 1/2$ is caused by the tetragonal distortion in the octahedral symmetry (Cr-O) that increase the crystal field factor Δ, like is shows in **Figure 6(A)** [9–12]. These results are compatibles with the photoluminescence results found by Yanez et. al. for the same compound [8].

The four features of the spectrum are explained with two *B* and B^* oblate axial signals typical by powder with $g_{\perp} > g_{\parallel}$ and hyperfine interaction $a_0 = A$. The parameters are summarized in **Tables 1** and **2**, corresponds to a system with $S = 1/2$.

Sample	***B* Signal**			
	$g_{ù}$	g||	g_{iso}	*A* (*mT*)
0	—	—	—	—
1	1.9731	1.9441	1.9634	4.93
2	1.9720	1.9462	1.9634	4.75
5	1.9726	1.9424	1.9625	5.89

Sample	B Signal			
	$g_{\perp}$	$g_{\parallel}$	g_{iso}	A (mT)
4 (300 K)	1.9762	1.9424	1.9649	5.89
4 (77 K)	1.9733	1.9448	1.9651	5.73

Table 1.
EPR parameters for B signal.

Sample	B^* Signal			
	$g_{\perp}$	$g_{\parallel}$	g_{iso}	A (mT)
0	1.9355	1.9194	1.9301	3.26
1	1.9352	1.9257	1.9310	2.97
2	1.9360	1.9204	1.9308	2.86
5	1.9350	1.9189	1.9296	3.10
4 (300 K)	1.9384	1.9189	1.9319	3.57
4 (77 K)	1.9372	1.9181	1.9308	3.33

Table 2.
EPR parameters for B^* signal.

For each B and B^* signal corresponds one Cr^{3+}, with spin $S = 1/2$, which is substituting to Ti^{4+} or Zr^{4+} in B sites into the octahedral symmetry with tetragonal distortion (high symmetry) slightly different one to other [27]. One of these cells correspond to octahedron with Cr^{3+} belong to crystalline cells localized into the ferroelectric grains, and the other belongs to crystalline cells on surface of the material grains. This interpretation is consistent with the interpretation published of the B and B^* sites distinguished of Mn^{4+} in $PbTiO_3$ [27].

6.2 Microwave power variation

The microwave power was varied from 1mW to 40mW for the samples and there is no change for EPR spectrum. The intensity of all features of the spectrum increase due power increase without differentiation. No distortion of the line shape of spectra is detected either, so until 40 mW power there is no sample saturation [9–18]. Qualitatively the 1, 2 and 5 samples present the same spectrum, but the quantitatively the EPR parameters change, **Tables 1** and **2**.

In addition to signals B and B^* the spectrum of sample 4 shows the isotropic G and H signals located at $g_H = 2.1449$ and $g_G = 2.0755$ respectively. By having these signals g values greater than 2.00 but around to zone to $g = 2.00$ and because they are anisotropic could be identified like two of three expected fine lines for Cr^{3+}, $s = 3/2$, it present the Zeeman split in a little crystalline field [5, 10, 14]. The positive deviation sign of value 2.0036 is explained for considerable difference between excited energy levels in ground state by Lx, Ly and Lz operators [12–18].

The signals B and B^* in the EPR spectrum for sample 4 at 77 K no changes respect to spectrum at 300 K, **Tables 1** and **2** but in the region for G and H signal at 77 K a third isotropic line with $g = 2.0716$ is resolved. The power variation study no shows changes line shape or line number or saturation effects at 300 K from 1 to 40 mW for this sample.

7. Conclusions

The ferroelectric $Pb_{0.95}Sr_{0.05}(Zr_{0.53}Ti_{0.47})O_3 + x\%wtCr_2O_3$ with $x = 0.0, 0.1, 0.2, 0.4$ and 0.5 samples produce a characterized spectrum well defined by absorption peaks. The signals B and B^* are present in the EPR spectrum for all samples doped with Cr and they no have saturation features for microwave power variation from 1 mW to 40 mW at 300K and 77K. Both B and B^* signals correspond to Cr^{3+} ion, with spin $S = 1/2$ which is substituting to Ti^{4+} or/and Zr^{4+} ions at B sites. One of signals corresponds to octahedron with Cr^{3+} ions belong to crystalline cells with tetragonal distortion localize into the grains. The other signal corresponds to octahedrons crystalline cells localize cells in the frontiers between the material grains, i.e., the EPR spectra shows two distinguishable B sites, which have been registered by B and B^* signals. The two B sites are similarly to B and B^* sites of Mn^{4+} in $PbTiO_3$ reporter in reference [27]. The microwave power increase does not increase the signals intensity separately, this means that the spectrum is not a superposition of two or more signals corresponding to different paramagnetic centers. The line width and the g-values are not changing with temperature or microwave power variation. The sample 4 shows the same spectrum for 0, 1, 2 and 5 samples. However, the sample 4 shows additionally the H and G signals forming a "fine triple" structure which indicates the Cr^{3+} presence at weak crystalline field environment in to local octahedral symmetry.

Author details

Veronica Lucero Villegas Rueda
Basic Sciences Department, Professional Interdisciplinary Unit of Engineering and Advanced Technology - IPN, Mexico City, Mexico

*Address all correspondence to: veyarle@gmail.com

References

[1] Scott J. F., Applications of Modern Ferroelectrics. *Science.*2007; 5: 954–959. DOI: 10.1126/science.1129564

[2] Irzaman H., Ferroelectrics and Their Applications. IntechOpen, 2018. DOI: 10.5772/intechopen.73075

[3] Blázquez A., García A and Carrascosa M. Biological applications of ferroelectric materials. Applied Physics Reviews 2018; 5: 041101; DOI:10.1063/1.5044472

[4] Kim T. Y., Kim S. K. and Kim S. W., Application of ferroelectric materials for improving output power of energy harvesters. Nano Convergence 2018 Dec; 5: 30. DOI: 10.1186/s40580-018-0163-0

[5] Calderón P. F., Oscilaciones Electromecánicas y pérdidas en cerámicasferroeléctricas [Thesis]. La Habana Cuba. Facultad de Física-IMRE, Habana University; 1999.

[6] Ramírez R. D., Estudio EPR de la reducción del Mn4+ a Mn2+ en el PbTiO3 Modificado [Thesis]. Ciudad de México, ESFM-National Polytechnic Institute; 2004.

[7] Wersing W., Libitz K. and Mahaupt J., Anisotropic Piezoelectric effect in modified PbTiO3 ceramics, IEEE Transactions on Ultrasonics ferroelectrics and frequency control; 36 (4):424–433, 1989.

[8] Hernández M., Durruty D., Costa J., Calderón F., Guerra J. and Yañez J. M., Photoluminescence in Pb0.95Sr0.05 (Zr1-xTix)1-yCryO3 ferroelectric ceramic system. Journal of Applied Physics 2014; 116: 043510. DOI:10.1063/1.4885762

[9] Drago R. S. Physical Methods for Chemistry. 2nd Ed. Surfside Scientific Publishers 1992. 660 p.

[10] Burns R. Mineralogical Applications of Crystal Field Theory, Cambridge Topics in Mineral Physics and Chemistry. Cambridge University Press 1964. 480 p.

[11] Roessler M. and Salvadori E. Principles and applications of EPR spectroscopy in the chemical sciences. Chemistry Society Review, 2018; 47: 2534. DOI: 10.1039/c6cs00565a

[12] Weil J. and Bolton J. Electron Paramagnetic Resonance: Elementary theory and practical application. Wiley, New Jersey 2007. 687 p.

[13] Zamorano R., El uso de EPR en la caracterización de compuestos con metales de transición, Academia Mexicana de Química Inorgánica, A.C. y CINVESTAV-IPN, México 4–21, 1993.

[14] Gordy W. Theory and Applications of Electron Spin Resonance, Duke University, John Wile & Sons 1980.

[15] Carrintong A. and Mclanchlan A. D., Introduction to Magnetic Resonance, Harper and Row, New York 1979.

[16] Pilbrow M. Transition Ion Electron Paramagnetic Resonance, Clarendon Press, Oxford, 1990.

[17] Feher G. Electron Paramagnetic Resonance with Applications to Selected Problems in Biology, Gordon and Breanch Science Piblishers 1969.

[18] Zadrozny J. M., Greer S. M., Hill S. and Freedman D., A flexible iron(II) complex in which zero-field splitting is resistant to structural variation. Chemical Science 2015. DOI: 10.1039/c5sc02477c

[19] Álvarez G., Caracterización de la absorción de potencias en materiales ferroeléctricos a muy altas frecuencias

[Thesis], ESFM-National Polytechnic Institute., México DF. 2006.

[20] Ramírez D. y Zamorano Ulloa R., Obtención Experimental de Señales en Resonancia Paramagnética Electrónica, Reporte Interno ESFM/F08/93, ESFM-National Polytechnic Institute, México 1993.

[21] Ramírez D., Resonancia de Espín Electrónica del compuesto (Mn(Pidieno (H2O)2)Cl 4H2O [Thesis]. ESFM-National Polytechnic Institute, México 1994.

[22] Jeol, Instructions ESR Spectrometer, Ed. Jeol LTD, Tokio, Japan, 1992.

[23] Basurto E. y Zamorano R., Simulación de espectros en el Software ESPRIT, Reporte Interno ESFM/F06/93, ESFM- National Polytechnic Institute, México 1993.

[24] Keeble D.J., Poindexter E. H. and Gerardo G.J., Electron Paramagnetic Resonance Studies of impurity defects in PbTiO3, Appl. Spectroscopy, 51(1); 1997:117–122

[25] Lian Xing He, et al, Effects of Cr2O3 addition on the piezoelectric properties and microstructure of PbZrxTiy(Mg1/3Nb2/3)1−x−yO3 ceramics J. of European Ceramic Society 2001:703–709. DOI:10.1016/S0955-2219(00)00256-9

[26] Mabbs F.E. and Collins D., Electron Paramagnetic Resonance of d Transition Metal Compounds, Elsevier, Amsterdam, 1992.

[27] Ramírez Rosales D., Zamorano Ulloa R., Electron spin study of the conversion of Mn4+ to Mn2+ in the Pb1-xEuxTi1-yMnyO3 ceramic system, Solid Stated Commun. 118; 2001: 371–376. DOI: 10.1016/S0038-1098(01)00072-2